*f*P

Also by Daniel Tammet

Born on a Blue Day

Embracing the Wide Sky

A Tour Across the Horizons of the Mind

Daniel Tammet

FREE PRESS

New York London Toronto Sydney

Free Press
A Division of Simon & Schuster Inc.
1230 Avenue of the Americas
New York, NY 10020

Copyright © 2009 by Daniel Tammet

Excerpts from *The Man Who Mistook His Wife for a Hat
and Other Clinical Tales* by Oliver Sacks reprinted with the permission
of Simon & Schuster Adult Publishing Group. Copyright © 1970,
1981, 1983, 1984, 1985 by Oliver Sacks.

First Free Press hardcover edition January 2009

FREE PRESS and colophon are trademarks of
Simon & Schuster, Inc.

For information about special discounts for bulk purchases,
please contact Simon & Schuster Special Sales at 1-800-456-6798
or business@simonandschuster.com

Title page photograph © 2008 by Kevin Beauchamp

Book design by Ellen R. Sasahara

Manufactured in the United States of America

1   3   5   7   9   10   8   6   4   2

Library of Congress Cataloging-in-Publication Data

Tammet, Daniel.
Embracing the wide sky: a tour across the horizons of the mind /
Daniel Tammet.
p. cm.
Includes bibliographical references and index.
1. Brain—Popular works. 2. Intellect—Popular works.
3. Savants (Savant syndrome) 4. Mnemonics. I. Title.
QP376.T36 2009
612.8'2—d22
2008030551

ISBN-13: 978-1-4165-6969-5
ISBN-10:   1-4165-6969-3

*This book is dedicated to the beauty found
in every kind of mind.*

Contents

Introduction

"How did you do that?"

"Sorry?"

"How did you do that?"

The scientist was looking at me with a puzzled expression. We were not in any laboratory, nor was he asking me about any of my memory, linguistic, or numerical skills. We were standing on a lawn outside the research center where I had come earlier in the day for a variety of cognitive tests. Next to him was my mother, who had accompanied me on the trip from London. We were in the process of having our photo taken together, when after a few moments in front of the camera I relaxed and started to step away. How, the scientist wanted to know, had I been able to perceive the photo being taken when, standing right next to me, he had not heard a click or seen any flash. Was my brain really that extraordinary?

Well, yes, but not for the reasons that the scientist imagined. Though the camera had indeed made no noise when the photo was taken, it *had* produced a pinprick of blurry red light. My autistic mind—wired in such a way that I am able to spot tiny details that most other people often miss—had perceived it effortlessly. After I explained this to the scientist, he asked for another photo to be taken. By looking carefully where I told him I had seen the red dot of light appear, he was able to see it, too.

For the record, I will confirm that I have no telepathic relationship with cameras, nor any extrasensory perception for knowing when a photo has or has not been taken. Rather, what I had done that day was simply an extreme form of an everyday act: to see. We rely heavily on our eyes to provide much of the information we obtain about the world around us, and it is for this reason that a significant portion of the human brain is devoted entirely to visual processing.

The scientist who thought I had perceived the photo being taken with the aid of some unknown power had arrived at a wrong but surprisingly common conclusion: that individuals with very different minds must use them in some fundamentally different, almost magical way. As one of the world's few well-known autistic savants, I have received all manner of strange requests: from being asked to predict the following week's winning lottery numbers, to requests for advice on building a perpetual motion machine. Little wonder then that conditions such as autism and savant syndrome remain poorly understood by most people, including many experts.

It is not only savant minds that are considered somehow supernaturally gifted and therefore set apart from those of most other people: the success of outstanding individuals in numerous fields, from Mozart and Einstein to Garry Kasparov and Bill Gates, has been attributed by many to minds they regard as unearthly and inexplicable. I think this view is not only erroneous but harmful, too, because it separates the achievements of talented individuals from their humanity; an injustice both to them and to everyone else.

Every brain is amazing. Researchers know this after many years of studying the minds of highly gifted people, as well as those of housewives, cab drivers, and many others from all walks of life. As a result, today, we have a far richer, more sophisticated understanding of human ability and potential than ever before. Anyone with the passion and dedication necessary to master a field or subject can succeed in it. Genius, in all its forms, is not due to any mere quirk of the brain; it is the result of far more chaotic, dynamic, and essentially human qualities such as perseverance, imagination, intuition, and even love. Such an understanding of the human mind enriches, rather than detracts from, the popular appreciation of the accomplishments of highly successful individuals.

This book is about the mind—its nature and abilities. It combines some of the latest neuroscientific research with my personal reflections and detailed descriptions of my abilities and experiences. My primary intention in writing it is to show that differently functioning minds such as mine (or Gates's or Kasparov's) are not so strange, in fact, and that anyone can learn from them. Along the way, I hope to clear up many misconceptions about the nature of savant abilities and what it means to be intelligent or gifted.

Chapter 1 looks at the fascinating complexity of the human brain and surveys some of the latest research findings from the field of neuroscience. Here I tackle head on some of the most common misconceptions concerning the brain, such as the idea that it does not change after birth or that the computer is a good analogy for how our brains work. I also assess several claims about savants and give evidence that indicates that savant brains are not so different from anyone else's.

Chapter 2 is a study of intelligence that questions whether IQ is an accurate indicator of intelligent behavior and looks at alternative ways of thinking about intelligence. I also examine the nature of genius and whether it is the result of innate talent, practice, or both.

Chapters 3, 4, and 5 include detailed descriptions of my own abilities in memory, language, and number sense respectively—areas where my autism helps me to excel. These chapters represent the most comprehensive personal account of savant ability ever written. Rather than encourage readers to merely gawk at the abilities of savants such as myself, I show that anyone can learn from them how to better understand and use his own mind.

Drawing again from my own personal experiences (as well as

those of other autistic individuals), chapter 6 explores creativity and the possibility that some neurological conditions predispose individuals to extraordinary forms of creative thought and perception. I describe little-known forms of creativity, such as the phenomenon of languages created spontaneously by some children, and refute the myth that autistic savants are incapable of genuine creativity, using examples from my own and others' work.

In chapter 7, I examine what the latest scientific research tells us about the complexity and limitations of our perceptions. I also explore how biological differences can cause different people to see the world in very different ways. Sections on the puzzle of optical illusions and the psychology of art demonstrate the malleability and subjectivity of our minds' eyes.

In chapter 8, I look at the nature of information and its relationship with our minds in the internet age of Wikipedia, twenty-four-hour rolling news broadcasts, and the ubiquity of modern advertising. I explore the role of words in shaping how we perceive and think about something, and how we share knowledge through such means as gossip and urban myths. I also give suggestions on how we can learn to navigate our information-dense world and reduce our risk of information overload.

In chapter 9, I demonstrate and explain the benefits of and methods for thinking mathematically. I show how ordinary intuitions can often lead to wrong conclusions, and how a lack of understanding of probability can result in bad choices. I also analyze complex real-world entities, such as lotteries and voting systems, from a mathematical perspective and show how certain statistical arguments for popular claims do not add up. A final section helps you learn how to use numbers and logic to think more carefully and successfully.

The tenth and concluding chapter looks at the future of the human mind, from the remarkable medical and technological breakthroughs that are transforming the treatment of injured and diseased brains, to the new insights of cognitive researchers that suggest our minds extend far beyond the confines of the head. I also assess the claims of futurists who assert that, inevitably, mind and machine will merge and give rise to a new "cyborg" species. I finish with some personal reflections on what I hope the future will bring for every kind of mind.

A final note: The title of this book was inspired by one of my favorite poems, a meditation on the mind by the celebrated nineteenth-century American poet Emily Dickinson. Every schoolchild should learn these verses:

> The brain is wider than the sky,
> For, put them side by side,
> The one the other will contain
> With ease, and you beside.
>
> The brain is deeper than the sea,
> For, hold them, blue to blue,
> The one the other will absorb,
> As sponges, buckets do.
>
> The brain is just the weight of God,
> For, heft them, pound for pound,
> And they will differ, if they do,
> As syllable from sound.

1

Wider Than the Sky

Our minds are miracles—immensely intricate webs of gossamer light inside our heads that shape our very sense of self and our understanding of the world around us. Moment by moment throughout our lifetime, our brains hum with the work of making meaning: weaving together many thousands of threads of information into all manner of thoughts, feelings, memories, and ideas. It is these processes of thinking, learning, and remembering that make each of us truly human.

And yet much of what goes on between our ears remains a mystery. Perhaps this is not surprising, considering that the brain is the most complex object known to man. Every action, from wriggling our toes to performing calculus, involves a breathtakingly sophisticated choreography of neural activity that scientists are only just beginning to understand. By adulthood, this jellylike mass of tissue weighs a little more than a kilogram, and contains around 100 billion neurons and as many as 1 quadrillion (1,000,000,000,000,000) connections—a greater number than stars in the known universe.

This unique complexity is a headache to researchers attempting to come to grips with it. Imagine the challenges inherent in trying to study something as intangible as a thought or a flash of inspiration. In spite of such problems, and the fact that the field is still in its infancy, neuroscientists have revolutionized our understanding of the brain in recent years, helping to treat a host of previously intractable illnesses and transform how we think about ourselves. I, for one, owe my life and self-understanding to such advances.

My brain has been scanned on numerous occasions, by doctors treating the epilepsy I had as a young child and more recently by researchers looking inside my head for clues to how my brain works—and for what it might tell them about how brains function generally. Having a brain scan is an unusual experience, beginning with a person in a white coat asking you whether you have any metal plates in your head or shrapnel in your body. This is because the scanner, known as an MRI (magnetic resonance imaging), uses an extremely powerful magnet to realign the atoms in your head so that they produce signals that a computer can process to generate a three-dimensional representation of the brain.

The scanner itself is the shape of a large cylindrical tube surrounded by a circular magnet. You lie on a movable table that slides into the center of the magnet. Being inside the scanner can make you feel rather claustrophobic, which is exacerbated by the need to remain completely still so that the imaging can work properly. The scanner is also very noisy, producing loud thumping and humming noises during the imaging. Fortunately, the entire examination usually takes less than an hour to complete, and is broken up into multiple runs (sequences) that each last several minutes.

The last time I was inside the scanner, a screen positioned above my head showed strings of numbers flashing across it that I was told to memorize. This task causes increased metabolic activity—including expanding blood vessels, chemical changes, and the delivery of extra oxygen—in the areas of my brain involved in my numerical skills. In a room next door, the scientists used computers to record this neural activity in highly detailed images that showed how my brain reacts to numbers. They could also visually compare my brain's activity with that of other people performing the same task.

It might all sound like something out of *Star Trek*, but this sort of technology is becoming increasingly commonplace around the world. Scientists are beginning to catch up to the complexity of the human brain, gaining insights that just a few years ago would have been unthinkable. In the pages ahead, I survey some of the most exciting findings and explore just what this new science teaches us about how all human brains work. First, let's look briefly at the dynamic stages in the development of the brain.

A Brief History of the Brain

I am not the same person I was ten or twenty years ago. That is because my brain is not the same as it was one or two decades ago, or indeed the same as it was one or two days ago. Our brains are in a constant state of flux that continues throughout each lifetime. This ongoing process of change and adaptation begins at the very dawn of our lives.

Each child's birth is a Big Bang—the dawning of a tiny yet extremely complex cerebral cosmos. Indeed, the process of creation is already apace within the first weeks of gestation, with the neurons forming at a dizzying rate: a quarter of a million per minute. A fetus's brain will produce around twice as many nerve cells as it will eventually need—nature's way of giving the newborn the best chance of coming into the world with a healthy brain. Most of the excess neurons are then shed midway through the pregnancy.

While the newborn baby's brain contains an immense wealth of neurons, they are still immature and many are not yet "wired together." Almost immediately following birth, the newborn's brain begins forming trillions of connections between the neurons; enabling the infant to see, hear, smell, think, and learn. These connections between different neurons (known as synapses) are formed by electrical activity inside the brain that is triggered by the child's experiences as he starts to absorb the sights, sounds, and sensations of the outside world.

By age two, the infant's brain has twice the number of synapses and uses twice the physical energy as an adult's brain. Around this time, many of these connections are pruned from lack of use while others are strengthened as the brain gradually fine-tunes itself. A person's neural architecture is largely fixed during these critical first few years of life.

Further rapid, significant changes in the brain occur during adolescence, making it in many ways the most tumultuous period of development since emerging from the womb. For example, researchers have found that the amount of gray matter (neural tissue) in the frontal lobe—the part of the brain where emotions, impulses, and judgments are processed—grows suddenly just before puberty, around age eleven for girls and age twelve for boys, followed by a pruning back throughout the teenage years. This process of frontal lobe development lasts well into the early twenties, which helps explain the impulsive behavior and mood swings of many teenagers.

Supporting the idea that the adolescent brain is a work in progress is another study, which asked a group of adults and teenagers to identify an emotion from pictures of people showing various facial expressions. The adults scored well, but many of the teenagers got it wrong. By scanning the participants' brains while they were taking the test, the researchers discovered that the teenagers were using a different part of their brain—the amygdala, the cerebral source of raw emotions and instinctive "gut" reactions—than the adults. The good news for parents is that the focus of brain activity gradually shifts from the amygdala to the frontal lobes as teens mature into adulthood.

Of course, age does not necessarily bring with it much opportunity for calm, reflective reasoning, for our most productive adult years are often marked by periods of chronic stress, which take a toll on the brain. In stressful situations, our bodies produce a heightened flow of a class of steroid called glucocorticoids, which make us more alert. Unfortunately, they also turn out to be toxic for the brain. As stress persists, neurons weaken and the hippocampus—a part of the brain crucial for learning and memory—starts to shrink.

Researchers have observed the same changes in the brain of adults suffering from depression, the most common form of mental disorder, a condition that affects as many as one in five people at some point in life. Scientists now know that antidepressants are effective in treating clinical depression not, as once thought, because they elevate serotonin levels in the brain, but because they help increase the production of a class of protein, known as trophic factors, which makes neurons grow.

As early as the 1960s, scientists had discovered that animals' brains were capable of generating new cells. In a series of pioneering studies on birds, neuroscientist Fernando Nottebohm showed that neurogenesis—the process of creating new brain cells—played an essential role in birdsong. Without these brand new neurons, male birds would be incapable of singing as they do. As much as 1 percent of the neurons in the birds' song center are created anew each day.

Not until the late 1990s, however, did scientists uncover evidence that adult neurogenesis occurs in humans, overturning the prevailing view that mature brains only lose cells, not grow new ones. More recently, researchers tracked some of these new neurons in test subjects, measuring the cells' electrophysiological activity to help determine their behavior. The team found that there is a two-week window, around a month following the new cells' birth, during which they act like the neurons of a newborn baby. The fact that these adult-born neurons can form connections indistinguishable from neurons that form early in life means that future scientists may be able to find a way to repair damaged brain tissue.

Adult neurogenesis occurs only in specific structures of the brain, such as the hippocampus. Elsewhere the numbers of brain cells decline as we age. On average, the brain loses between 5 percent and 10 percent of its weight between the ages of twenty and ninety. Despite these facts, age-related cognitive decline is not inevitable. Many people work, learn, and study even into advanced old age. A fine example is the poet Stanley Kunitz, who became poet laureate of the United States at ninety-five and published his final book at age one hundred. The normal aging process may even confer unique advantages on the brain that form the basis for wisdom. Research suggests, for example, that advancing age increases emotional stability, because the brain increases its control over negative emotions as it becomes more able to draw on positive ones. As older people accumulate more knowledge, they also build more networks of connections in the brain, helping it to work better.

Changing Our Minds

Neuroscience's breakthrough discovery of the brain's ability to grow and change throughout our lifetime, known as neuroplasticity, contradicts the classical view of the adult brain as inflexible and mechanical, each part having a fixed, specific role, ticking along monotonously, and gradually wearing down with age like a machine. In its place, we find a new model of the adult brain as a supple, dynamic organ capable of responding successfully to injury and even of thinking itself into new synaptic formations. The implications are staggering, not only for patients with neurological injury or disease but for everyone.

Take one example of our new understanding of the brain's resilience: doctors once considered damage caused by a stroke irreversible, but recent advances in treating the condition indicate otherwise. Doctors now consider stroke a "brain attack" and treat it along the lines of a heart attack, with various medications and physical and mental exercises that aid functional recovery by working with the brain's natural plasticity. A common result of stroke, for instance, is damage to the brain's motor cortex, which helps control and execute our body's movements. Doctors now use a treatment called constraint-induced (CI) movement therapy that works by encouraging additional areas of the cortex to take over the role of the damaged part. In CI therapy, the patient is forced to use the affected arm by restraining the other in a sling. This helps the stroke patient overcome a phenomenon known as "learned nonuse." Obviously, when an area of the brain loses function, the part of the body linked to this area is also affected and loses mobility. Unable to move the affected limb, the patient compensates by using the other. In time the brain adapts so that recovery of movement becomes possible, but by that time the patient has already "learned" that the limb is no longer functional.

In CI therapy, the patient uses the affected arm intensively for two weeks, in regular activities such as dressing, eating, cooking, and writing. At the same time, the patient undergoes a rigorous six-hour daily program of physiotherapy. The increased use of the affected arm stimulates the area of the brain connected to it, with the result that the cortex assigns new neurons to moving the arm. In time, the patient is often able to recover much of his limb's former ability.

An example of how even our perceptions can alter the structure of the brain comes from the phenomenon of phantom limbs: the sensation that an amputated limb is still attached to the body and can move and feel normally. Most amputees experience these often painful phantom sensations. Some report that the missing limb feels as if it takes on a life of its own, beyond their control. The neurologist V. S. Ramachandran hypothesized that phantom limbs were the result of "cross-wiring" in the somatosensory cortex, a part of the brain that is activated whenever the body is touched. In fact, the complete surface part of a person's body is mapped on that of his brain, so that if, for example, someone touches his hand, neurons in the corresponding part of the brain respond. Professor Ramachandran's theory was inspired by the work of researchers who discovered that the brain modified its sensory mapping when part of it stopped receiving impulses. Following an amputation, the neighboring areas in the cortex (those correlated with the arm and face) take over the role of the area for the missing hand.

To test his hypothesis, Ramachandran worked with an amputee named Tom who had lost his left forearm in a car accident and subsequently reported an itching sensation and pain in his phantom fingers. Ramachandran blindfolded Tom before using a cotton bud to stroke various parts of Tom's body, asking him where he felt the sensations. When Ramachandran touched Tom's cheek, Tom felt sensation in his missing thumb; when he touched Tom's upper lip, Tom felt the touch in his phantom index finger. Touching Tom's lower jaw caused feeling in his missing little finger. In this way, Ramachandran was able to find a complete map of Tom's missing hand on his face. Subsequent brain scans of patients with phantom limb syndrome confirmed the professor's findings: the brain adapted its inner structures to maintain a feeling that the body was whole.

Ramachandran surmised that the phantom limb pain felt by some amputee patients was the result of a form of learned paralysis—similar to the learned nonuse of limbs seen in stroke patients. Using this idea, Ramachandran developed an original method for alleviating this persistent pain: a "mirror box" that helps the patient to "unlearn" the paralysis by tricking his brain into thinking the missing arm is still there.

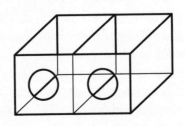

The patient places his good arm into one of the box's holes and the amputated one into the other. As though clapping his hands, the patient performs repeated movements in the direction of a mirror running down the center of the box. The reflected image of the good hand moving makes it appear as though the missing hand is also moving. The patient uses this illusion to "exercise" the phantom hand and release it from its paralysis. Repeated use of the box has brought long-term relief to some patients.

Our brains can rewire themselves based on our experiences. This realization raises an interesting question: can we use our brains' plasticity to enhance our senses and even create new ones? Yes, says cognitive scientist Peter König, inventor of the feelSpace belt. Lined with thirteen vibrating pads, the wide belt uses an electronic compass to detect the earth's magnetic field. With each step the user takes, the vibrator that points nearest to magnetic north starts to buzz. In time, the wearer is able to orient himself. One subject who tried out the belt for six weeks developed an intuitive map of his city inside his head and eventually felt as if he could never get lost; even in a completely new place he could always find his way home.

Direction is not something humans can detect innately, though some birds, bats, fish, and turtles can. For example, female sea turtles migrate across the Atlantic Ocean and are able to retrace their route back to the beach where they were born to lay their eggs, an epic journey of 8,000 miles. Researchers believe the turtles "read" the earth's magnetic field to help them navigate the currents that sweep the ocean. König's feelSpace belt suggests that we humans, too, could learn to navigate like sea turtles. Some researchers say it is conceivable that we could one day develop other senses found in the animal kingdom, such as the infrared vision of snakes and piranhas or the ultrasonic hearing of bats and dolphins.

The possibility that our senses could be adapted to meet particular needs drives a team of neuroscientists in Wisconsin who specialize in building sensory prosthetics to help people with balance and vision problems. Among their creations is a mouthpiece fitted with 144 tiny electrodes attached to a pulse generator, which produces an electric current against the tongue. It works by generating a vibrating square in the center of the user's tongue that responds to his movements (for example, moving to the left side if the person turns left) to help him maintain his balance. The device has proved especially helpful in the treatment of patients with damaged inner ears, with the effects lasting for hours or even days following its removal.

The mouthpiece has also helped individuals with visual impairment by teaching them to use their tongue as a surrogate eye. Images picked up by a camera are translated into patterns of electric pulses that trigger touch-sensitive nerves on the tongue. Users say they perceive the stimulation as three-dimensional shapes and features. They also develop some ability to navigate their surroundings. In one experiment a blind man was able to walk through woods and find his wife. Though the main emphasis of the team's research has been rehabilitation, they also envisage the technology being used by people who do not have sensory deficits. For example, infrared cameras could inform soldiers of the position of enemy troops by stimulating different parts of their tongue.

What about the rest of us? Might it be possible to draw on the brain's extraordinary plasticity without the use of hi-tech gadgetry? Absolutely. Take, for example, the following experiment that looked at the brain's ability simply to think itself into new configurations: Harvard neuroscientist Alvaro Pascual-Leone taught a group of nonmusician volunteers to play a simple five-finger piano exercise, then had them practice in the lab for two hours a day for five days. After a week, brain scans of the volunteers showed an increase in the amount of territory the brain devotes to moving the fingers. Pascual-Leone then repeated the experiment with another group of volunteers, asking them this time to rehearse the same five-finger sequence in their heads, while holding their hands still and imagining how they would move their fingers. Subsequent scans of these volunteers showed the same result as for those who had actually played the sequence with fingers. The scientists' conclusion: mental imagery may be just as good as actual practice.

Sports psychologists utilize this principle to help competitors improve their performance. The process of visualization helps the muscles learn the proper way of moving through a complex activity (such as a golf swing). Studies confirm that players who use visualization in their training outperform those who only practice physically.

That the brain can change as a result of the thoughts we think has significant implications for our health. We see this most clearly in the development of therapies that can modify a person's neuronal connections, alleviating mental illness and even growing our capacity for happiness. Cognitive behavioral therapy (CBT) for example, aims to modify or eliminate unwanted thoughts and change behavior in response to them. Research has shown that it can be just as effective as medication in treating anxiety-related conditions. Brain scans indicate that CBT works by decreasing activity in the frontal cortex while increasing it in the limbic system, the brain's emotional center. In so doing, the therapy reshapes how the person processes information: decreasing rumination and helping him to swap negative thinking patterns for more positive ones.

If thinking about thinking can effect such positive changes in the brain, what about its most popular form: meditation? The neuroscientist Richard Davidson, at the University of Wisconsin, organized a study involving two groups of practitioners. On one side were eight Buddhist "adepts" (monks who had each accumulated at least ten thousand hours of meditation), and on the other ten college students who had undergone a crash course in how to meditate. In a basement lab, each participant was wired up to an electroencephalograph that measured his brainwave activity as he performed a form of meditation known as "nonreferential compassion"—opening up to feelings of unlimited love and generosity towards all living beings. The scientists monitoring the EEG noticed a surge in one kind of brain wave in particular, gamma, which is involved in perception and the consolidation of information. The increase in these normally weak and transient waves suggested that the participants were engaged in particularly intense and focused thought. Even more remarkable was the fact that the highly experienced meditating monks produced gamma waves thirty times stronger than those of the novice students.

Professor Davidson also used functional magnetic resonance imaging (fMRI) to identify the areas of the monks' and students' brains that were active during their compassion meditation. All the subjects' brains showed activity in regions involved with positive emotions, while those that keep track of what is self and what is other became quieter. But there were also interesting differences between the monks and novices. The monks' brains showed much greater activity in regions of the brain associated with empathy and maternal love. And when the monks were generating feelings of compassion, activity in the part of the brain associated with negative feelings was swamped by activity in the area correlated with happiness. In contrast, the students' brains showed no such activity.

In each instance, the monks with the most hours of meditation showed the most significant brain changes, supporting the idea that mental training can make the brain more prone to states of happiness, compassion, and empathy. According to this evidence, being happy is a skill that you can train yourself to learn. That is certainly one beautiful thought worth meditating on.

Seizures and Savants

The modern revolution in our understanding of the brain offers new hope for those suffering from previously incurable neurological disorders. At the same time, it is producing unprecedented insights into conditions once clouded by ignorance and prejudice. Just as the twentieth century saw a transformation in our knowledge of outer space, the twenty-first promises the same for our individual inner ones.

The treatment of epilepsy—the result of sudden, usually brief, excessive electrical discharges within the brain—is a good example of this progress. Once considered a reaction to a supernatural force or the phases of the moon, epilepsy can now be treated with drugs that are effective in eliminating seizures in 70 percent of cases. In the future, computational modeling technology that currently predicts complex natural systems, such as earthquakes, tidal waves, and hurricanes, might be used to help predict when seizures will occur in patients with epilepsy. Researchers in the United States are pioneering brain implants that use computer chips to predict and prevent seizure activity.

Such advances are also changing the general public's negative perceptions of minds considered different from the norm. For instance, since the mid-nineties, scientists have recognized the sheer diversity within the autistic experience: from the silent, rocking child to the brilliant, socially awkward scientist. We now know that there are as many forms of autism as there are individuals with the condition. Public awareness of autism's complexity is also on the increase, helped in large part by the mainstream success of works such as Mark Haddon's novel *The Curious Incident of the Dog in the Night-Time*, as well as by the rising profile of talented, high-functioning autistic individuals themselves.

In my 2006 memoir, *Born on a Blue Day*, I describe my own experience of growing up with Asperger's, a relatively mild and high-functioning form of autism. Though I exhibited many of the most common traits of autism in childhood—social isolation, difficulties with abstract thought ("seeing the bigger picture"), and communication problems—as an adult I lead a successful, happy, and independent life with a career, relationship, and numerous friends and intellectual pursuits.

I am happy to say that my success in learning to overcome many of the limitations of my autism is not unique. Indeed, it is becoming increasingly clear that having an autism spectrum disorder need not necessarily be any barrier to both personal and professional achievement. Fellow notable "Aspergians" include: Richard Borcherds, professor of mathematics at the University of California, Berkeley, and a winner of the prestigious Fields Medal—the mathematics equivalent of the Nobel Prize; Professor Vernon L. Smith, winner of the Nobel Prize in economics; Bram Cohen, author of the BitTorrent computer downloading protocol; Dawn Prince-Hughes, PhD, a primate anthropologist and primatologist; and Satoshi Tajiri, the creator of Pokémon. In spite of such varied accounts of high achievement by autistic individuals, the "typical" person with autism is still widely misperceived as being severely disabled, antisocial, and obsessed by exclusively trivial, impractical interests. But this is a cruel stereotype. The truth is that there is no one "typical" form of autism; every autistic person is different.

Perhaps the main reason why this stereotype persists is because of autism's association with Savant syndrome, another complex and little-understood neurological condition. Indeed, many people's knowledge of autism is still largely or entirely derived from its depiction in the Oscar-winning 1988 movie, *Rain Man*, starring Dustin Hoffman as a gifted but considerably disabled autistic savant. Ever since, savants in books and movies have been frequently depicted as brilliant but flawed individuals, oppressed under the sheer weight of their gifts. However, the science that inspired the film's makers, and subsequently captivated its audience, is more than twenty years old, long past its sell-by date. The modern understanding of autistic savants is, fortunately, much more sophisticated and compassionate.

When scientists first informed me, four years ago at age twenty-five, that I met the diagnostic criteria for Savant syndrome, I could not help but think back to the cinematic Raymond Babbitt character. How could I—an otherwise healthy young man with a partner, job, and friends—be considered a "rain man"? With research I soon discovered that there was much more to being a savant than a single Hollywood portrayal.

Savants are defined as individuals with a developmental disorder (usually, but not always, autism) who possess extraordinary abilities related to their condition in one or more fields. I am considered a "prodigious savant," which means that my abilities would be considered exceptional even if they occurred in someone without any developmental disability. There are estimated to be fewer than fifty prodigious savants living worldwide.

A version of savant syndrome was described in the medical literature as far back as 1789, when Dr. Benjamin Rush—regarded as the "father of American psychiatry"—described the calculating ability of Thomas Fuller, who "could comprehend scarcely anything, theoretical or practical, more complex than counting." When asked how many seconds there were in a year and a half, Fuller answered in about two minutes: 47,304,000. A century later, in 1887, Dr. J. Langdon Down used the word "savant"(from the French *savoir,* "to know") for the first time to describe ten cases of individuals who displayed striking abilities alongside a range of developmental problems. One of the men described built large, intricately detailed model ships from hand-fashioned parts, while another case involved a boy who could recite large parts of the six volumes of *The Rise and Fall of the Roman Empire* from memory.

Having such a rare, distinctive condition, savants have long been subject to all manner of speculation, misunderstanding, and—sadly, at times—exploitation. This is especially regrettable because, as I know from my own experience, high-functioning autistic savants are fully capable of complex human emotions and of making meaningful contributions to society. Savant abilities are the result of imaginative human minds, not dry mechanical processes.

Counting Matchsticks

Among the many popular misconceptions of savants is that their abilities are somehow supernatural, beyond the scope of scientific study. In fact, researchers have been examining savant abilities for decades, and their findings have been published and reviewed many times. My own abilities have been studied by neurologists at laboratories in the United Kingdom and in the United States and have been the subject of several published scientific papers.

Probably the most famous and influential study of savant abilities was an informal one by the American psychiatrist Oliver Sacks, written up in his 1985 book, *The Man Who Mistook His Wife for a Hat*. Unfortunately, Sacks's account of autistic savant twins, cited in many journals, mainstream articles, and popular science books, is the source of a number of persistent misconceptions concerning autistic savants and the nature of their abilities. Here, for example, is his account of an incident that was subsequently dramatized in the *Rain Man* movie:

A box of matches on their table fell, and discharged its contents on the floor: "111," they both cried simultaneously; and then, in a murmur, John said, "37." Michael repeated this; John said it a third time and stopped. I counted the matches—it took me some time—and there were 111. "How could you count the matches so quickly?" I asked. "We didn't count," they said. "We saw the 111." ... "And why did you murmur "37", and repeat it three times?" I asked the twins. They said in unison, "37, 37, 37, 111" ... "How did you work that out?" I said ... They indicated, as best they could ... that they did not "Work it out," but just "saw" it ... is it possible ... that they can somehow "see" the properties ... as qualities, felt, sensuous, in some immediate, concrete way?

The twins' ability to visualize numbers and their properties sounds very much like my own. Indeed, 111 is an eminently visual number to me: full of beautiful bright white light. However, Sacks makes no mention of where the box of matches came from in the first place. He does say that the matches were on the twins' table, suggesting that they were already in the twins' possession when Sacks visited them for the first time. Also noteworthy is Sacks's failure to indicate the twins' line of view of the matchsticks as they tumbled to the ground. This is important because falling groups of objects do not remain separate and distinguishable as they descend: a stream of more than one hundred falling matches would certainly involve some of the matches being obscured by the others. Bearing these facts in mind, a much likelier explanation of what Sacks saw is that the twins had previously counted the matches in the box on their table and knew how many were inside. It is even possible that they had chosen how many to place inside the box originally. After all, 111 is a particularly beautiful (and matchstick-like) number—one the twins might well have considered something of a collector's item. What we can say for sure is that the claimed ability to instantly discern large quantities has never been reported in any other savant, nor has any scientific study been able to confirm it.

Another widely reported claim in Sacks's account of the twins has been just as influential in shaping the public's misconception of savant abilities:

They were seated in a corner together . . . locked in a singular, purely numerical, converse. John would say a number—a six-figure number. Michael would catch the number, nod, smile, and seem to savor it. Then he, in turn, would say another six-figure number, and now it was John who received, and appreciated it richly . . . What on earth was going on? I could make nothing of it . . . I contented myself with noting down the numbers they uttered . . . All the numbers, the six-figure numbers, which the twins had exchanged, were primes . . . I returned to the ward the next day, carrying the precious book of primes with me . . . I quietly joined them . . . After a few minutes I decided to join in, and ventured a number, an eight-figure prime . . . simultaneously, they both broke into smiles . . . Then John . . . thought for a very long time . . . and brought out a nine-figure number . . . his twin Michael responded with a similar one . . . I, in my turn, after a surreptitious look in my book, added . . . a ten-figure prime I found in my book . . . John, after a prodigious internal contemplation, brought out a twelve-figure number. I had no way of checking this . . . because my own book . . . did not go beyond ten-figure primes . . . an hour later the twins were swapping twenty-figure primes, at least I assume this was so, for I had no way of checking it.

An article published in 2006 by Makoto Yamaguchi, an educational psychologist, in the *Journal of Autism and Developmental Disorders,* questions this account. Yamaguchi points out that a book of every prime number up to ten digits would have to list more than 400 million numbers; something that is impossible to do in a single book with a reasonable font size. Sacks replied that his book and other resources are now lost, but acknowledged that the book may have included only smaller numbers. Other leading researchers, such as the mathematician Stanislas Dehaene and neuroscientist Brian Butterworth, have also expressed skepticism about Sacks's report.

Like the matches incident, Sacks's description of savants generating twenty-digit primes is unparalleled—no other such account exists in any of the literature about savants and their abilities. In contrast, all the scientific research done to date on savants' prime-number skills (including my own) indicates a typical range of between three and five digits, though it is possible for savants to use their knowledge of primes to make judgments of numbers in excess of five digits that are better than average. Without Sacks's original notes, it is impossible to say for certain how the twins generated their numbers, or even whether the numbers given were all correctly identified as prime numbers. Contrary to another widespread misconception, even savants make mistakes.

Perhaps the most unfortunate fallacy to arise from the Sacks account is that savants are freakish and alien:

> "[The twins] are . . . a sort of grotesque Tweedledee and Tweedledum . . . they are undersized, with disturbing disproportions in head and hands, high-arched palates, high-arched feet, monotonous squeaky voices, a variety of peculiar tics and mannerisms . . . glasses so thick that their eyes seem distorted, giving them the appearance of absurd little professors, peering and pointing, with a misplaced, obsessed and absurd concentration . . . like pantomime puppets to start spontaneously on one of their "routines."

The best one can say of this description of the twins is that it is distinctly unsympathetic. It is also starkly at odds with the reality of my own life and those of numerous other modern high-functioning savants. Matt Savage, for example, an American teenage autistic savant musician, has released several albums, won many awards, and toured the world. The British artist savant Stephen Wiltshire (now in his thirties) was awarded a Member of the Order of the British Empire by the Queen in 2007 for services to art; he has since opened his own art gallery in London where he displays his drawings, accepts commissions, and meets his fans. Gilles Tréhin, thirty-five, from Nice, France, has created a vast, complex city inside his head. His drawings and detailed notes about "Urville" have been published in several languages. Tréhin has a long-term girlfriend—

also a high-functioning autistic—with whom he speaks at conferences to improve public understanding of autism.

Evidence that savant talents are rooted in natural (if extraordinary) brain processing comes from research carried out by the Australian scientist Professor Allan Snyder, director of the Centre for the Mind in Sydney. Autistic thought is not incompatible with the "ordinary" kind, Snyder argues, but is rather a variation of it—a more extreme example.

To test his theory, Snyder and his colleagues used a technique called transcranial magnetic stimulation (TMS), which involves sending a series of electromagnetic pulses via electrodes into the subject's frontal lobes—the idea being to shut down temporarily the left hemisphere of the brain in order to boost the right side (the side most implicated in savant skills). As expected, a quarter hour of stimulation boosted artistic and proofreading abilities in several nonautistic subjects. The newfound skills rapidly disappeared as the subjects' brains returned to their prior state.

Dancing with Numbers

Snyder's experiment suggests that the gap between savant and "normal" minds is not as great as once thought. Its results confirm that a savant's brain activity is much closer to that of an "average" person's than to the functioning of a supercomputer—a still commonly employed analogy for how savants' minds work. Perhaps this is why I have never liked the term "human computer," as often used in books and newspaper articles to describe savant skills. As I will show in chapter 5, while computers crunch numbers, I dance with them.

The computer analogy is not limited to savants—a number of scientists and philosophers have used it to describe the functioning of the human brain in general. Such attempts to explain the brain by reference to something vaguely analogous are, of course, nothing new. Through history, the brain has been compared to a water clock, a steam engine, and a telephone switchboard, among other things. The modern image of the human brain as nothing more than "a computer made of meat" (to use the expression of computer scientist Marvin Minsky) is, in my view, just as impoverished and reductionist. Like the analogies that came before it, the "brain as computer" model neither explains savant abilities nor those of anyone else.

For decades, cognitive scientists—inspired by the comparison between brains and computers—believed that the mind behaved in a strictly feed-forward direction, passing discrete packets of information from one part of the brain to the next, like a computer processing data. Recent experiments suggest otherwise, however, adding to the growing scientific consensus that the brain is a much more dynamic system than previously thought.

In a 2005 study by psychologists at Cornell University, forty-two students were asked to respond to audio cues by clicking on pictures of different objects on a computer screen. When the students heard a word like *candle*, and were shown two pictures of objects that did not sound alike (such as a candle and a jacket), the trajectories of their on-screen mouse movements were generally straight and direct to the candle. But when the students heard "candle" and were presented with two pictures of objects with similar sounding names—like *candle* and *candy*—they were slower to click on the correct object, and their mouse trajectories were much more curved.

The researchers conclude that the students were processing what they heard even before the whole word was spoken. But when the two words closely resembled each other, the students—unable to tell immediately which picture was correct—considered both choices simultaneously. Rather than immediately move the mouse to one picture, and then correct their movement if they realized they were wrong, the students allowed the mouse to wander in an "intermediate gray area" between the two images, waiting until the ambiguity was resolved.

These scientists contend that when we think of cognition as nonlinear and dynamic, it is possible to be in two different brain states at the same time before arriving at the final interpretation. This view contrasts sharply with that in which computers are either in one state or another, moving in lockstep from one to the next.

The fundamental dissimilarity of minds and machines is further illustrated when we compare their respective strengths and weaknesses. For example, while most people struggle to add up their restaurant bill, even the most modest household laptop can solve immense sums in mere fractions of a second. As far as calculating power goes, people are no match for today's computers. An impressive demonstration of this made headlines in July 2007, when computer scientists at the University of Alberta in Canada announced that they had solved the game of checkers after nearly two decades of research. Their program, called Chinook, can never lose; the best any opponent can achieve against it, with perfect play, is a draw. With 500 billion billion theoretically possible board positions, it is the most complex game solved to date. The scientists used hundreds of computers over several years to run through game after game in order to calculate the sequences that would lead to a win, a loss, or a draw. Eventually, the Chinook program gathered so much information that it could determine the best move to play in every situation.

Another game at which computers undoubtedly excel is chess, described by the German poet Goethe as "the touchstone of the intellect." In 1997, IBM's Deep Blue computer defeated the reigning world chess champion Garry Kasparov in a six-game match played on the thirty-fifth floor of a Manhattan skyscraper. What I found most remarkable about the match, however, was that Kasparov was able to compete at all (he lost by a single point, 3.5 to 2.5, after winning a match the previous year 4 to 2) against a machine that could calculate at an estimated speed of 200 million moves per second and search up to twenty moves ahead.

What Kasparov lacked in brute calculating force, his brain made up for in intuition and highly developed pattern-recognition skills. A century of psychological research has shown that chess grandmasters like Kasparov use a rapid, knowledge-guided perception of the board to make very good moves after just a few seconds thought. Former world champion José Raúl Capablanca described this perceptive, intuitive way of thinking when he said, "I see only one move ahead, but it is always the correct one."

Adriaan de Groot, a Dutch psychologist, demonstrated the enhanced perceptual abilities of chess grandmasters in a 1966 study by comparing their skills to those of novice players. De Groot showed the players a complex, midgame chess position for five seconds and then asked them to reproduce the board from memory. He discovered that the grandmasters were able to reconstruct the board nearly perfectly, whereas the novices could reproduce only a few of the pieces' positions. More recent studies suggest that chess grandmasters have more than a million patterns in memory to draw upon during play.

That knowledgeable players can outwit computers at their own games is not news to aficionados of the ancient Chinese game of Go. Even the most powerful programs developed to play Go have so far been baffled by its deceptive simplicity. Unlike chess, Go is played using identical stones (black for one player and white for the other), which the players take turns placing across a 19-x-19 grid board. The object of the game is to stake out the largest territory possible by carefully placing the stones and surrounding your opponent's forces in order to eliminate them.

What makes Go so much more difficult for computers than chess is its underlying computational complexity. In a game of chess, the player faces an average of around 35 possible moves to consider at each turn. In Go, the figure is nearer 200. After four moves, the number of possible board positions in chess is $35 \times 35 \times 35 \times 35 = 1,500,625$, whereas in Go the number is:

200 x 200 x 200 x 200 = 1,600,000,000—more than a thousand times greater. Looking ahead 14 moves (as the most powerful chess computers can) in Go would involve calculating through ten thousand trillion possibilities; even a computer program as fast as Deep Blue would take a year and a half to consider a single move.

Worse for computers is the fact that the immensely intricate positions in Go are incredibly difficult for a program to evaluate, because the very nature of the game resists simple analysis. The game has no single piece like a king whose loss decides the winner. Even the amount of territory each player has captured at any given point is not a good indication of the likely outcome, as the placement of a single stone by an experienced player has the potential to turn a large tract of an opponent's territory into his own in an instant. This fuzziness at the heart of the game is what gives human players the edge. Go experts employ heightened pattern-reading skills, spatial awareness, intuition, and empathy (to sense their opponent's likely strategy or tactics)—the very hallmarks of human intelligence that even the fastest and most powerful computers are unable to replicate. Some experts believe that it will be at least a hundred years before a program is developed that can give the Go masters a good game.

Games like Go are not the only activities in which humans can still easily outperform computers. Take the ability to have a conversation with another person, for example, something that most people (including high-functioning savants) can do well, even effortlessly, but with which computers struggle a great deal. Indeed, the British mathematician Alan Turing proposed in 1950 that computers be considered truly intelligent only if they could converse like a person, the so-called Turing test. Since 1990, the Loebner Prize has awarded medals annually to the inventors of computer programs—known as chatterbots—that can maintain the most humanlike dialogue. The most recent winner was "Joan," designed by the British entrepreneur Rollo

Carpenter, which has more than 5 million lines of conversation available to it. But Joan's ability is still nowhere close to passing Turing's famous test, as the following short conversation between the program and me demonstrates:

Daniel: Hi, Joan, my name is Daniel.

Joan: Hi, Daniel.

Daniel: Where are you from?

Joan: Hastings, originally.

Daniel: Oh, great! I don't live far from there.

Joan: Loughborough.

Daniel: How do you pronounce that?

Joan: Well, just right forward, S-t-i-g.

Daniel: Sorry?

Joan: Do you know their origin?

Daniel: Whose origin?

Joan: I'll ask again. Do you know anything about human relationships?

Daniel: I thought we were talking about Loughborough. What's that got to do with human relationships?

Joan: Our current conversation.

Daniel: What do you do for a living?

Joan: I work at a bar to pay the bills; but I'm trying to become a professional writer.

Daniel: I'm a writer, too. What sort of stuff have you written so far?

Joan: Lots of stuff; I've told you some of it.

Daniel: No, you haven't.

Joan: It seems to me that your RAM refresh cycle is too long.

Daniel: Do you write about computers?

Joan: No, it isn't. The movie is about the Civil War through the eyes of a spoiled rich girl.

Daniel: Which civil war?

Joan: I am not talking about civil war, my friend.

The irony here is that, like many other individuals with an autism spectrum disorder, I have had to learn over many years how to converse well with other people. Even to this day, "small talk" does not come naturally to me. Yet even I can see through the Joan program immediately. Perhaps the Turing test researchers would have more success if they modeled their future programs on autistic individuals who, like me, have learned to navigate the intricacies of human conversation.

How Does the Human Brain Learn?

If even the most powerful, sophisticated modern computers are unable to handle quotidian tasks such as holding a conversation, how then do our brains learn to do such things, and often to do them well? This question has both fascinated and frustrated neuroscientists for decades. Jeff Hawkins, director of the Redwood Neuroscience Institute in California, believes that the human brain is like a memory system that records everything that happens to us and makes intelligent predictions based on those experiences. Unlike computers, which must be programmed to solve problems, our brains are self-learning—they automatically build a model of the world over time through the senses. Knowledge is stored hierarchically in the brain's neocortex—the area of the brain responsible for almost all high-level thought and perception. For example, the memory of what a cat looks like is not stored in one location. Instead, low-level visual details such as fur, paws, and ears are stored in lower cortical regions while high-level structures—such as the head or torso—are stored in the higher regions.

The main advantage of this hierarchical system is that it allows us to reuse knowledge, transferring what we have learned about one thing to help us learn about something new. For example, once a child has learned what a cat is, he needs much less time and effort to learn what a dog is, because cats and dogs share many low-level features such as fur, paws, and tails that he does not need to relearn every time he sees a new animal. However, this reusing of formerly learned knowledge can also result in misunderstanding—for example, when a child thinks that the fraction 1/9 is greater than 1/7 because he has previously learned that 9 is more than 7.

Knowing lots of facts is clearly not enough to develop competence in an area of study. A deeper understanding of the subject is needed in order to transform facts into usable knowledge. In chess, for instance, learning to play well involves far more than knowing the names of the different pieces and how they move around the board; understanding concepts such as piece development, central control, space, pawn structure, and king safety is essential for acquiring expertise in the game. Chess grandmasters can remember complex midgame positions at a glance by seeing patterns and relationships between the pieces that novice players do not. Years of study have refined their understanding of the game so that what is relevant is perceived immediately and what is not is simply ignored or not even seen.

Obtaining guidance from experts in the subject you are learning helps you to both develop the conceptual framework necessary to master a field of study and avoid many of the potential discouraging pitfalls that can lead to discouragement. Getting feedback on wrong answers allows the brain to eliminate them from its bank of possible future responses. Consequently, learners can ignore poor choices and focus their attention on good ones. For this reason, learning activities that feature high levels of feedback (both positive and negative), such as cooperative games, and peer review and discussion, are much more likely to be effective than passive ones.

Emotions and motivation are also fundamental to learning. Strong emotion connected with an experience helps the brain store information in a way that is more accessible and more easily retrieved. Too much stress, on the other hand, can result in reduced blood flow to the frontal lobes, impairing the ability to think and remember clearly. Taking pleasure in a task is an especially good way to learn well. Our brains release a neurotrans-mitter called dopamine in anticipation of the pleasure we expect to derive from a particular activity. The dopamine motivates us, increasing our energy and drive and encouraging us to engage in the activity. If our brain's expectation of pleasure in a cer-tain activity is met, dopamine levels remain elevated. If the plea-sure enjoyed is even greater than predicted, dopamine levels are increased and we engage even more persistently in the activity. Conversely, if the activity is less pleasurable than anticipated, dopamine levels drop sharply.

Though practice falls short in its proverbial claim to make us perfect, it is necessary if we want to obtain long-term or permanent results from our study. As we practice, our ability to perform a task grows; the rate and shape of improvement are relatively stable across different tasks—the so-called learn-ing curve or power law of practice. This law is ubiquitous: from short perceptual or cognitive tasks—such as reading inverted text or performing mental arithmetic—to team-based longer-term ones—such as manufacturing machine tools or building ships—the rate that people improve with practice follows a simi-lar pattern.

The learning curve shows us that, while practice will always help improve performance, the most dramatic improvements happen first, with diminishing returns thereafter. It also implies that with sufficient practice individuals can achieve comparable levels of performance in a wide range of tasks, but only if the learner does not relax as soon as an acceptable performance is reached. Rather, expertise comes solely from a continuous pro-cess of structured, diligent study.

Ultimately, though, it is what we learn, more than how, that helps determine the shape of our lives and even the kind of peo-ple we become. For this reason, how we use our minds remains a personal choice we each have to make. After all, what our brains help give us, more than anything else, is our own uniqueness and the myriad tastes and talents that emerge from it. What we do with them, and how, is part of the adventure of becoming ourselves—a unique, personal process that we cannot shortcut, nor try to bend to the expectations of others. Quite simply, no one way of thinking or learning is superior to another. Just as there is no single definition of the life well lived, there cannot be one of the mind well used. As Nobel Prize–winning physicist Richard Feynman put it: "You have no responsibility to live up to what other people think you ought to accomplish. I have no responsibility to be like they expect me to be: it's their mistake, not my failing."

2

Measuring Minds: Intelligence and Talent

What we do with our brains is usually described as our "intelligence," but what exactly is intelligence? I am not smart enough to answer that. Perhaps that is because there is no one agreed upon definition of what it means to be intelligent. In this the concept shares much with another notoriously intangible one, that of love. The French philosopher Michel Onfray alludes to this problem in his maxim: "There is no such thing as love, there are only the proofs of love." Similarly, many scientists consider that the concept of intelligence exists meaningfully only in its concrete expression, such as that supposedly measured by IQ tests. I am not so sure. As with love, attempts to reduce intelligence to a particular explanation or set of criteria have failed time and time again. In fact, I believe the inverse of Onfray's famous dictum to be nearer the mark: there is no such thing as proofs of intelligence, only intelligence.

I have seen firsthand the problems inherent in trying to define something as personal and complex as intelligence according to any generalized theory or formula. As a child my behavior was often limited, repetitive, and antisocial—far from what most people would consider intelligent. Indeed, I struggled during my first years at school because of the effects of medication I was given to control my epilepsy, and because my style of imaginative and idiosyncratic thinking did not suit the standard rituals of "one size fits all" schooling. Without access to specialist support or "gifted child" programs, my teachers did not know what to make of me and had little choice but to leave me to my own devices.

This is the typical story line for many wasted minds, brimming with talent but starved of opportunities to exercise it. Fortunately, I did not follow the script: with the continuous support and encouragement of my family, I found ways to feed my gifts, from frequent visits to local libraries and play acting with my siblings, to games of Scrabble and poetry recitations with the few friends I was lucky enough to make. As my confidence and social skills improved, so did my ability to articulate the beautiful mental landscapes of words, numbers, and ideas inside my head. Eventually, I flourished academically, even as I struggled in other respects, and little by little I learned to trust in my mind's ability to do wonderful things. I also realized that my differentness was a blessing, not the burden I had long feared it to be, and found it possible to accept myself for how I was and am.

If intelligence is, as I contend, a concept too subtle and nebulous to "prove" in any scientific way, is IQ testing necessary or appropriate? To help me answer this question, I decided to undergo the process myself—the first time I have ever had my "IQ score" assessed. Supervised by a qualified educational psychologist, the test comprised a series of one-on-one tasks taken from the most commonly used evaluation of adult intelligence—the Wechsler Adult Intelligence Scale (WAIS), named after its creator, American psychologist David Wechsler. It was Wechsler who first assigned the arbitrary value of 100 to score a person's intelligence—adding or subtracting points depending on whether the test subject did better or worse than average on his range of tasks.

On a graph, the distribution of IQ scores of a representative sample of test takers follows the familiar bell-shaped (or Gaussian, after the mathematician Carl Friedrich Gauss) curve, with half the scores falling above 100 and half below. According to the graph, 68 percent of scores will fall between 85 and 115 (technically described as being within 1 standard deviation, or SD, of the mean) and 95 percent of scores between 70 and 130 (within 2 standard deviations). Scores below 70 or above 130 are very rare and therefore difficult to measure accurately.

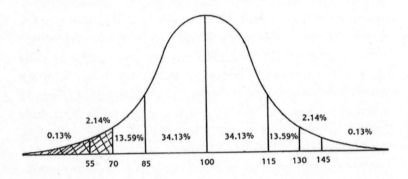

Test givers assign the following, rather dry and limited, assessments of intelligence according to the subject's score:

130 and above: Very Superior
120–129: Superior
110–119: High Average
90–109: Average
80–89: Low Average
70–79: Borderline
69 and below: Extremely Low

The test is divided into two main sections: verbal and performance, each of which has seven subtests. These evaluate a range of cognitive skills including verbal comprehension, perceptual organization, working memory, and processing speed. Each subtest takes around ten minutes to complete, starting with the easiest questions and becoming gradually more difficult. For example, on the test of digit span, the subject is first asked to memorize and recite back a sequence of 4 digits forwards and backwards, which increases progressively up to sequences of 9 digits. Similarly, the general knowledge subtest begins with a question like: "What is the capital of France?" and culminates with ones like: "Who wrote the Aeneid?" and "What is the circumference of the earth?" The different tests are administered one after another, without break, over two to three hours.

I was particularly struck during my own test by how banal the various tasks were, from being asked to put a series of pictures into a sequence that tells a story, to detecting the presence or absence of a target symbol in lines of different symbols. None of the tasks required the ability to reflect on a topic or critically analyze an idea, nor was there any opportunity for creative thought. Worse still was the fact that many of the questions had only a single anticipated response—any other given, however imaginative or poetic, was considered incorrect. For example, how would you answer the following question: "What do a fly and a tree have in common?" Little surprise that these restricted response questions were the ones with which I struggled most.

In spite of such difficulties, the tasks played to many of my strengths, such as memory, vocabulary, arithmetic, and general knowledge, and my WAIS IQ score was calculated as 150 (equivalent to a score of 180 on the Mensa Cattell test, which uses a different method to assign a final score). In view of the result, the psychologist suggested I apply for membership in the Mensa Society—the world's largest, oldest, and best-known high-IQ society. Rather like a gentlemen's club for bright people, membership is limited to those with an IQ score in the top 2 percent of the population (WAIS 130+, Mensa Cattell 148+).

Mensa has been criticized as elitist and too preoccupied with puzzles and games by those who would prefer it to be a forum for truly intelligent debate and discussion. Even one of its founders, Dr. Lancelot Ware, voiced this concern during the society's fiftieth anniversary in 1996: "I do get disappointed that so many members spend so much time solving puzzles." For this and other reasons, I have no intention of joining Mensa myself. To quote Groucho Marx: "I don't want to belong to any club that will have me as a member!"

Having an IQ score has not in any way altered how I think about myself. Though I accept that such tasks are a useful means of measuring some specific forms of cognitive ability (such as when doctors need to assess the effects of a head injury), I am wary of considering it in any way descriptive of the person I am or an oracle for the kind of life I will lead. That said, the peculiar experience of undergoing an IQ test did stir my curiosity about its origins, and what I discovered was a story more mind-boggling than I could ever have imagined.

Tipping the Scales: The History of IQ

The first scientific attempt to study human intelligence began in the eighteenth and nineteenth centuries with craniometry, the "science" based on the premise that skull size was correlated with intelligence. The larger the skull, the larger the brain, and the larger the brain, the higher the intelligence. Chief among the craniometricians was the American scientist Samuel George Morton, who had amassed a collection of more than a thousand skulls by the time of his death in 1851. Morton measured the skulls by filling them with mustard seed or lead shot, pouring the contents back into a graduated cylinder to obtain a reading of the skull's volume in cubic inches. He published his figures in several books, purporting to show that intelligence could be divided along racial lines, with white Europeans at the top and black Africans at the bottom.

At a time when it was customary for many eminent public figures to bequeath their brains to science after death, the supporters of craniometry looked to the weighing scales for proof of their claims. Some of the luminaries did very well—against the European average of 1,300 to 1,400 grams, the German poet Schiller's weighed in at 1,785 grams and the Russian author Turgenev's a whopping 2,012 grams. Others flatly contradicted the scientists' expectations, however: Walt Whitman's brain was a distinctly average 1,282 grams, the French novelist Anatole France's a mere 1,017 grams. Scientist Paul Broca, an outspoken advocate of craniometry as a reliable measurement of human intelligence, tried to account for these small brains by suggesting that their owners had either been very old, small, or that their brains had been poorly preserved.

Such views were extremely influential in their time. Maria Montessori, the Italian education reformer, measured the circumference of children's heads in her schools and concluded that those with bigger heads learned faster and had better prospects. Alfred Binet, director of the psychology laboratory at the Sorbonne and inventor of the first IQ tests, also measured the heads of pupils at various schools, but failed to find any significant difference between the head sizes of children designated by teachers as their smartest and stupidest. Binet subsequently described the idea of measuring intelligence by measuring heads as "ridiculous."

In 1904, Binet was commissioned by the education ministry in France to develop techniques for identifying children with special educational needs. He chose a series of everyday tasks, such as counting coins and arranging words into a sentence, that were administered individually by trained examiners. Binet then assigned an age level to each task and defined the youngest age at which a child of normal intelligence should be able to complete it. Each child taking the Binet test was first given the tasks for the youngest age, progressing in sequence until he could go no further. The age associated with the last task that the child was able to perform became his "mental age." In 1912 a German psychologist, Wilhelm Stern, invented the term "intelligence quotient" (IQ), to describe a person's intelligence numerically by dividing his mental age by his chronological one. For example, a child aged 10 with a mental age of 12 would have an IQ of 120 (12 divided by 10, x 100 to eliminate the decimal).

Binet's ideas for evaluating a child's mental age were introduced to America by Henry Goddard, research director at a school for "feeble-minded girls and boys" in New Jersey. Goddard translated the test into English and used it to evaluate people whom he considered mentally retarded. Considering intelligence to be hereditary, he believed it necessary to identify the "feeble-minded" in order to prevent them from having children. To this end he proposed that they be compulsorily sterilized or placed in institutions. Goddard later revised his views, but by then his work had already been widely published and had become influential in public opinion and policy-making.

A professor at Stanford University, Lewis Terman, subsequently modified and popularized what became the Stanford-Binet IQ test, the standard for many tests that followed it. Unlike Binet's original set of tasks, which were intended only to help identify children with special educational needs, Terman wanted every child to be tested and ranked according to innate ability. Like Goddard, Terman believed that his IQ test was necessary to bring those who performed poorly on it "under the surveillance and protection of society" so as to reduce "the reproduction of feeble-mindedness." He argued that this would lead to "the elimination of an enormous amount of crime, pauperism, and industrial inefficiency."

The use of IQ tests by the United States military during the First World War enhanced their credibility and visibility, and in the postwar years they grew in popularity, spawning a multimillion-dollar industry. Many companies started testing procedures to choose among job applicants and to identify employees' promotion prospects. Schools, too, developed various intelligence scoring programs. In Britain the "11 Plus" exam (so named because it was taken by every child in the country at age eleven), was introduced in the 1940s, determining whether the pupil subsequently studied at an academic or technical school—with clear consequences for the child's future career choices in adulthood.

The most pernicious effects of the belief in intelligence as a fixed, heritable entity were seen in the United States, where it influenced the passing of restrictive immigration laws based on the idea that all immigrants except those from Northern Europe were of "surprisingly low intelligence" (according to Goddard). Vast numbers of immigrants were deported in the early twentieth century as a result of failing an IQ test. In some states, the tests also helped to legitimize the forced sterilization of supposedly "defective" individuals.

The validity of IQ testing remains a lively, bitterly contested topic of debate to this day. An intense academic and media reaction against IQ was spurred by the 1994 publication of the provocative book, *The Bell Curve*, by Harvard professor Richard Herrnstein and political pundit Charles Murray, who misinterpreted data on intelligence to lead to some racist conclusions. Among the authors' controversial claims were that alleged differences in intelligence corresponded to race and social class, and that society is becoming increasingly "intellectually stratified," leading to the emergence of a "cognitive elite." Herrnstein and Murray's research on the nature of, and reasons for, large and widening economic and cultural disparities in modern society is voluminous (some 900 pages) but ultimately disappointing. For one thing, the authors focus too much on a single, narrow measure of intelligence, "g" ("general intelligence"), at the expense of considering the sheer diversity of human abilities. They also seem to conflate intelligence with level of education, though the two do not necessarily coincide: the list of highly talented individuals who did not do well at school is huge and varied, including notables such as Thomas Edison, Charlie Chaplin, Winston Churchill, Woody Allen, and Vincent van Gogh among many, many others. Herrnstein and Murray also relate economic success to intelligence, which is not a reliable indicator of an individual's talent: van Gogh famously sold just one of his paintings during his lifetime.

The Mismeasure of Man

From the earliest days of IQ testing, its own inventor Alfred Binet cautioned that the tests could not be seen as a meaningful measure of individual intelligence: "the [Binet] scale, properly speaking, does not permit the measure of intelligence, because intellectual qualities are not superposable, and therefore cannot be measured as linear surfaces are measured." Binet also criticized those who claimed that a person's intelligence could not be increased: "some recent thinkers . . . [have affirmed] that an individual's intelligence is a fixed quantity, a quantity that cannot be increased. We must protest and react against this brutal pessimism; we must try to demonstrate that it is founded on nothing."

The American journalist Walter Lippmann, another early critic of IQ tests, wrote in the 1920s:

> the intelligence test . . . is an instrument for classifying a group of people, rather than a "measure of intelligence." People are classified within a group according to their success in solving problems, which may or may not be tests of intelligence . . . The tests are all a good deal alike. They all derive from a common stock, and it is entirely possible that they measure only a certain kind of ability . . . We cannot measure intelligence when we have not defined it.

Such criticisms were updated and expanded in the 1981 book, *The Mismeasure of Man*, by American geologist and evolutionary biologist Stephen Jay Gould. In what remains to this day the most comprehensive critique of intelligence testing, Gould argues against what he describes as the main theme of biological determinism, that "worth can be assigned to individuals and groups by measuring intelligence as a single quantity." He considers the methods used by IQ researchers compromised by two fallacies: reification, "our tendency to convert abstract concepts into entities" (such as "IQ") and ranking, "our propensity for ordering complex information as a gradual ascending scale."

Gould devotes a significant portion of his book to an analysis of statistical correlation, which psychometrists often point to when arguing for the validity of IQ tests and the heritability of intelligence. They claim, for example, that an IQ test accurately measures a person's general intelligence, using data that show high correlation between a person's score and future success in life. IQ researchers also claim that the scores of testees who are closely related show higher correlation than those only distantly related. In response, Gould points out that correlation is not equivalent to cause. For example, Gould's age and the population of Mexico measured over time showed high positive correlation, but this does not mean that his age goes up *because* the Mexican population increases. In addition, a high positive correlation between a parent's IQ and that of her child is evidence for both the idea that IQ is genetically inherited, and that it is the result of shared social and environmental factors. The data can be used to argue either side of the case and is inconclusive at best.

A further weakness of IQ testing, according to its critics, is that the tests are written so that knowledge of the answers depends on a person's sex, class, and culture. Yet, in spite of the fact that there is enormous variation in individuals' educational backgrounds based on a wide range of factors, IQ scores for many tests are typically adjusted only for age. Many consider this unfair. For example, a child from a middle-class family will likely have better access to books and tutoring than a child of the same age from a poor family, yet the test scores for both children are treated identically.

To see how coming from a different culture can affect your IQ score, try answering the following test questions based on an Australian Aborigine culture, the Kuuk Thaayorre, in Far North Queensland:

1. What number comes next in the sequence: one, two, three?
2. Wallaby is to animal as cigarette is to?
3. Which items may be classified with sugar: honey, witchetty grub, flour, water lilies?
4. Sam, Ben, and Harry are sitting together. Sam faces Ben, and Ben gives him a cigarette. Harry sits quietly with his back to both Ben and Sam and contributes nothing to the animated conversation going on between Sam and Ben. One of the men is Ben's brother; the other is Ben's sister's child. Who is the nephew?
5. You are out in the bush with your wife and young children and you are all hungry. You have a rifle and bullets. You see three animals all within range—a young emu, a large kangaroo, and a small female wallaby. Which should you shoot for food?

Answers:

1. The answer is *mong* or "many"—the Kuuk Thaayorre system of counting only goes up to three.
2. "Tree" is the right answer. This comes from the Kuuk Thaayorre speakers' early experience of tobacco, which was "stick" tobacco; hence it is classified with trees.
3. All the items are classified with sugar as they all belong to the class of objects known as *may,* or "vegetable food." Witchetty grubs are so assigned because they are found in the roots of trees, and honey because it is also associated with trees and hence fruit. Flour is included in the *may* category because it resembles some of the community's own processed vegetable foods.
4. This one is easy for the Kuuk Thaayorre. An avoidance

taboo prevents a mother's brother and sister's son from talking face to face. Sam and Ben are obviously brothers because of their unrestrained interaction while Harry, with his back turned to both his uncles, is the respectful nephew.

5. The small female wallaby is the right answer. Emu is a food that may be consumed only by very old people. Kangaroos (especially large ones) may not be eaten by parents or their children; the children will get sick otherwise.

Some IQ tests can also be criticized on mathematical grounds. For example, Mensa's own IQ test uses multiple-choice questions (typically with five options to choose between per question), which means that luck can play a considerable role in determining an individual IQ score. Assuming that a person of "average" ability (for answering the kinds of questions typically found in an IQ test) knows the answers to 50 of a test's 100 questions, he can expect to score on average 60/100 (where he guesses on each of the remaining 50 questions he does not know the answer to, and scores an average of 10/50 because the odds for each question is 1 in 5). Someone else, also knowing the answers to 50 of the questions, might be luckier in his guesses and score 68/100. Conversely, a third person, knowing the answers to 50 questions like the first two, might be less lucky and score just 52/100.

In addition, the bell curve distribution for IQ scores tells us that two-thirds of the world's population have an IQ somewhere between 85 and 115. This means that some 4.5 billion people around the globe share just 31 numerical values ("He's a 94," "You're a 110," "I'm a 103"), equivalent to 150 million people worldwide sharing the same IQ score. This reminds me of astrology lumping everyone into one of twelve signs of the zodiac. Is human intelligence really so uniform that it can be summed up in just a handful of figures?

Multiple Intelligences

Clearly, something more than a single test score is needed to guide our evaluation of a person's intelligence. I believe it makes more sense to consider intelligence as a complex phenomenon that is best described as a synthesis of various skills and abilities. In this way a person can be considered intelligent in some respects and less so in others. For example, how would you evaluate the intelligence of the following individuals?

- A Nobel Prize winner who regularly forgets where he put his car keys

- A thrice-divorced chess champion

- A company chief executive with a history of stress-related heart problems

- A doctor who smokes and drinks heavily

- A brilliant musical composer plagued by creditors

In view of such contradictions, various theorists have sought to broaden the traditional understanding of intelligence. In the 1980s Yale psychologist Robert Sternberg proposed the "triarchic" (three-part) theory of intelligence, which states that intelligence consists of three main aspects: analytic intelligence (the ability to analyze, evaluate, and compare), creative intelligence (skill in using past experiences to achieve insight and deal with new situations), and practical intelligence (the ability to adapt to, select, and shape the real-world environment). "Successfully intelligent" people, according to Sternberg, are those who are aware of their particular strengths and weaknesses in the three areas of intelligence. They work out how to maximize their strengths, compensate for their weaknesses, and develop further their abilities in order to achieve success in the future.

Like Sternberg, Howard Gardner, a professor of education at Harvard University, believes that there is more than one kind of intelligence—eight, to be precise—with every person having a unique blend of each. His theory of "Multiple Intelligences" was made famous by his book, *Frames of Mind*, first published in 1983. Using a range of criteria, including development history, evolutionary plausibility, and support from experimental psychology tasks, Gardner identified these eight different intelligences:

Linguistic intelligence: involving both spoken and written language, the ability to learn languages, and the capacity to use language to achieve certain goals. Examples: writers, poets, lawyers, and speakers.

Logical-mathematical intelligence: the capacity to analyze problems, perform mathematical operations, and investigate issues scientifically. Examples: scientists, engineers, and mathematicians.

Musical intelligence: skill in the performance, composition, and appreciation of musical patterns. Musicians of all kinds are obvious examples of this intelligence.

Bodily-kinesthetic intelligence: using parts or the whole of one's body to solve problems. Examples: athletes, actors, and dancers.

Spatial intelligence: includes having a very good sense of direction, as well as the ability to visualize and mentally manipulate objects. Examples: artists, architects, and engineers.

Interpersonal intelligence: the capacity to understand the feelings, intentions, and motivations of other people. Examples: salespeople, politicians, and therapists.

Intrapersonal intelligence: the ability to understand oneself, one's feelings, goals, and motivations. Examples: philosophers, psychologists, and theologians.

Naturalistic intelligence: the ability to draw upon certain features of the environment, to grow and nurture new things, and to have a facility for interacting with animals. Examples: farmers, gardeners, and conservationists.

Many educators in the United States who have adopted Gardner's theory of multiple intelligences to use in their schools report improved exam results, parental participation, and classroom discipline. A Harvard-led study of forty-one schools supported using the theory, and reported that in these schools there was "a culture of hard work, respect, and caring; a faculty that collaborated and learned from each other; classrooms that engaged students through constrained but meaningful choices, and a sharp focus on enabling students to produce high-quality work."

A third theory that challenges the "IQ" conception of intelligence is that of "Emotional Intelligence," or EQ, first popularized by psychologist and science journalist Daniel Goleman's 1995 bestseller, *Emotional Intelligence: Why It Can Matter More Than IQ*. Goleman argues that a person's emotions play a significant role in thought, making decisions, and future success. He defines this form of intelligence as a set of skills that include impulse control, self-motivation, empathy, and the ability to relate well to others.

EQ is not considered the opposite of IQ: some people have lots of both, others little of either. Rather, researchers like Goleman are interested in how the two forms of intelligence complement one another; how, for example, a person's ability to handle stress affects concentration and the ability to think effectively.

Self-awareness, Goleman argues, is the key to being truly

emotionally intelligent, because it allows a person to exercise self-control. With sufficient self-awareness, it is possible to develop various coping mechanisms that allow a person to move from a negative emotional state to a more positive one: counting to ten as a means of letting the sensation of sudden anger subside, for example.

As with Gardner's theory of multiple intelligences, Goleman's EQ concept has been adopted by various schools in the United States, which use it to develop "emotional literacy" programs, aimed at helping students learn to manage their anger, frustration, and loneliness. Children who are angry or depressed are incapable of learning well, and those with long-running emotional difficulties are liable to drop out altogether. Improving these students' self-esteem and self-motivation helps them to perform better in exams.

Like IQ, each of these alternative conceptions of intelligence has been criticized. Critics of the "multiple intelligences" theory, for example, point to the lack of empirical evidence supporting it. Critics of the EQ concept argue that it measures conformity rather than ability: who, after all, is to say when a person's anger or sadness (or other emotion) is or is not appropriate to a particular situation? EQ's skeptics also point out that scientific studies have failed to find a convincing link between high self-esteem and better academic performance.

While mindful of such criticisms, I believe that there is considerable value in considering human intelligence in ways that allow us to appreciate the enormous diversity in how people think and behave.

Is Genius in the Genes?

Controversies over the nature of intelligence also extend to its origins: Are examples of remarkable talent or giftedness the result of nature or nurture (or both)? Public and scientific opinion is polarized on this question in good part because the stakes are so high, with the different views leading to starkly different social and political implications. If, for example, genes determine a person's talents, then there is little we can do to improve the minds and bodies we are born with. On the other hand, if environmental factors are what count, then education, encouragement, and access to opportunity are more important than who an individual's parents or grandparents are.

A possible answer to this conundrum emerged from a peculiar project undertaken by an eccentric American millionaire optometrist, Robert Klark Graham, in the late 1970s. By collecting sperm from Nobel laureates and distributing it to intelligent women, he hoped to create an entire generation of geniuses. Graham believed, like many of the early advocates of IQ testing, that society was under threat from "retrograde humans" who were breeding unchecked and swamping the intelligent. His sperm bank project—the Repository for Germinal Choice—aimed to produce super-smart people to help save the world from genetic decline. Graham set up the sperm bank using an underground bunker in the backyard of his ranch near San Diego, California, hoping it would be the first of many such banks all over the United States producing "creative, intelligent people who otherwise might not be born." He personally persuaded several Nobel laureates to donate and advertised for prospective mothers in a Mensa magazine.

The story of the repository broke in 1980 with a *Los Angeles Times* article that resulted in Graham being pilloried in the press, accused of attempting to create a "master race." Graham, however, was unrepentant, and his bank continued, producing its first baby in early 1982 followed by dozens, then hundreds more, until the bank was finally closed in 1999.

Most of the project's offspring remain anonymous, so it is impossible to know whether Graham's experiment worked. However, a few have come forward, such as Doron Blake—the repository's "poster boy"—now in his midtwenties. It is ironic that Blake, who is training to become a teacher, does not believe that geniuses can be created in the way that Graham envisaged: "It doesn't matter how a child is made in terms of the genes and chromosomes . . . it's how the child was raised and nurtured that really matters."

That viewpoint is certainly shared by Laszlo Polgar, a Hungarian educational psychologist and author of the book, *Bring Up Genius!*, which claims that any child can become a genius given the right environment and parental training from an early age. Polgar developed his theory by studying the biographies of hundreds of great intellectuals, among whom he identified a common theme: early and intensive specialization in a particular field. To prove the point, he decided to raise his own daughters—Susan, Sofia, and Judit—with the goal of producing a trio of chess champions.

Lazslo and his wife Klara were no ordinary parents—they battled the Hungarian authorities for permission to homeschool the girls and taught them various languages, including Esperanto, as well as high-level math, in addition to hours of daily chess training. The curiously rigid schedule even extended to twenty minutes each day being set aside for joke telling.

For all the eccentricities of their upbringing, the results are undoubtedly impressive: in January 1991 Susan became the first woman ever to achieve the grandmaster title; later that year Judit, at age fifteen, broke the former world champion Bobby Fischer's record to become the youngest grandmaster ever. All three sisters have won major tournaments over the years and beaten many of the world's best players (both male and female).

The success of the Polgar sisters appears to bolster the arguments put forward by developmental psychologist Michael Howe in his 1999 book, *Genius Explained*, who suggests that, rather than the result of any natural endowment, genius is purely the result of hard work, perseverance, and good luck. Howe also argues that genius should be defined by achievement rather than the possession of any innate quality, pointing out that "unsuccessful genius" is an oxymoron.

Howe supports his thesis with a number of biographical accounts of great minds, such as Mozart, Darwin, and the Brontë sisters. Howe speculates that Mozart had studied some 3,500 hours of music with his instructor father by his sixth birthday. Darwin was among the best prepared scientists of his generation: chosen at twenty-two to take part in the voyage of the *Beagle*, he was considered the most capable young biologist. The Brontës, too, are shown to have worked extremely hard towards their future achievements, perfecting their skills by spending hundreds of hours writing stories for each other from early childhood.

Another scientific supporter of the "genius made, not born" theory is professor K. Anders Ericsson, of Florida State University, who has spent more than two decades studying the achievements of experts in a range of fields. Ericsson's research suggests that great performances only follow sustained and systematic practice. In one study he found that classical pianists who performed at the highest levels had invested more than 10,000 hours in practice by their twentieth birthdays—a figure up to five times greater than that of their less successful fellow pianists. Similar results were found for other musicians, chess players, and athletes.

Clearly, lots of work is necessary for high achievement, but it is far from certain whether it is sufficient in and of itself. Many scientists are critical of what they call the "drudge theory" of genius. Among them is neuroscientist Ognjen Amidzic, who describes the example of the Polgar sisters as a "beautiful coincidence." Amidzic, who runs his own private laboratory in Switzerland, once aspired to become a professional chess player. He practiced constantly, even moving to Russia as a teenager to study intensively with grandmasters, but he reached a plateau in his early twenties and had to quit. His disappointment led him into cognitive science in an attempt to understand what went wrong.

Using brain scans, Amidzic discovered a big difference between grandmasters and highly trained amateurs like himself: grandmasters, he found, used far more of the frontal and parietal cortices of their brains (areas responsible for long-term memory and higher-level processing). Amateurs, by contrast, have more activation in the medial temporal lobes (responsible for short-term memory). Without the information they acquire with study being passed into their long-term memory, amateurs often end up having to relearn many of the things they have already learned. As a result, they are unable to progress significantly beyond a certain playing level.

Amidzic's research suggests that a person's chess processing ratio of frontal-and-parietal cortices to medial temporal lobes is genetically predetermined, remaining stable over time regardless of the amount of study or practice. In grandmasters, that ratio is 80–20, while for skilled amateurs like Amidzic it is generally around 50–50. Retrospective analyses of older players seem to confirm this, showing strong correspondences between their ratios and highest historical chess ratings. Amidzic even claims that by using these ratios he is able to predict the exact chess rating at which a child will eventually peak.

My own family provides further evidence that an individual's talents are natural as well as "nurtural." We were five girls and four boys born in one of the most economically deprived parts of southern Britain. Our childhoods were generally happy but unconventional. With so many children and so little money, my parents had neither the time nor the economic means to nourish our individual interests as they would have liked. Our success came despite the environment in which we were raised, not as a result of it. As examples, my sister Claire obtained an Education Achievement Award by coming in the top three in her school for exam results, and subsequently obtained a degree in English and philosophy (she is currently studying for a master's degree in archiving). Two other sisters, Maria and Natasha, passed their school exams at sixteen with straight A and A-plus grades, despite missing many of their classes due to ill health. My brother Steven, who like me has Asperger's, has taught himself how to play both the guitar and the Greek lute and is currently teaching himself Russian and Chinese. My youngest sibling, Shelley, was reading the novels of Jane Austen and the Brontë sisters for fun before the age of ten.

I think of talent according to its etymological roots: as a weight or inclination that pushes a person in a particular direction. Accordingly, I believe that everyone is born with certain talents, which dedication and hard work help to realize. I broadly agree with the scientific consensus that says that high achievement is the result both of genetic and environmental factors. In other words, I agree with the idea of biological variation, though not in biological determinism, about which I agree with the criticisms of Stephen Jay Gould, among others.

Attempts to suggest that a person's talents are purely the result of hard work are, in my opinion, misguided. Being captive to the largely random effects of our environment is hardly more humane a view than being captive to our genes. The truly humane view, and the scientific one, is surely in seeing talent as something that emerges spontaneously—like a bird in flight— from the subtle and complex interactions of a multitude of genetic and environmental elements. Without our individual talents, we would all be blank slates, slaves to whatever environment we were born into. Instead, every person can have confidence knowing that, by our very humanity, we each have something unique and beautiful to contribute to the world around us. In the end, it is not the size of our brains that matters, but the depth of our spirits.

3

Seeing What Is Not There

Imagine entering a room around which a dozen everyday objects are scattered. After a few minutes, you step outside while someone else enters and removes one of the items. When you return a short while later, you will likely be able to tell immediately which of the objects has been taken. As though endowed with some superhuman power, you will do this by seeing *what is not there*. Such is the magic of memory.

When I recited the mathematical constant Pi (3.141 . . .) from memory to 22,514 decimal places in March 2004, it seemed like magic to many people. In fact, the achievement (a European record) was the result of weeks of disciplined study aided by the unusual way in which my mind perceives numbers, as complex, multidimensional, colored, and textured shapes. Using these shapes, I was able to visualize and remember the digits of Pi in my mind's eye as a rolling numerical panorama, the beauty of which both fascinated and enchanted me.

One of my fondest memories from the Pi event in Oxford is the profound sense of joy I felt at that visual experience of the numbers' beauty. The public recitation of number after number after number developed into a kind of meditation for me, as I grew more and more wrapped up in their flow. Although the digits of Pi are, mathematically speaking, strictly random, my internal representation of them was anything but—filled with rhythmic strokes and structures of light, color, and personality. From this random assembly of digits I was able to compose something like a visual song that meandered through every contour of my mind, through which I was able to hear the music of the numbers.

The ability to remember vast amounts of very specific kinds of information is one common to all savants. Kim Peek, the inspiration for the Dustin Hoffman character in *Rain Man*, can remember huge quantities of facts and figures spanning more than a dozen subjects from the thousands of books he has read over the years. Kim and his father Fran travel all over the United States giving performances at schools, colleges, and hospitals in which Kim answers trivia questions from members of the public. According to Fran, Kim rarely gets stumped by any of the questions posed him.

When I met Kim and his father in Salt Lake City in the summer of 2004, as part of the filming for a documentary on my life with savant syndrome, I took the opportunity to perform a simple experiment to evaluate the precise nature of Kim's memory. The film crew accompanying me brought Kim a gift, a book on British history, and I asked him to read one of the pages from it at random for several minutes. Afterwards I selected two facts from the page Kim had just read; one (the birth date of a famous historical figure) that he would likely have read in many other history books; the other (a precise quantity of a material) was likely only found in that book. Kim easily recalled the first fact, but not the second.

The reason for this is that Kim's memory is not in fact photographic (the idea that a savant's brain can remember information like a camera is a myth). Instead, Kim remembers his vast repository of factual information by weaving the facts he learns together into a mental network of many thousands of different associations and interconnections. Speaking with Kim quickly illustrated this highly associative form of thinking (which my own experience suggests is typical of how savant minds really work)—a word or name spoken in the middle of a sentence would suddenly elicit a flood of loosely related other words, names, or facts. Kim would even sometimes respond to a word or fact by bursting into song—the lyrics of which might contain some minor link to what was just mentioned. The ability to weave a new piece of information into this dense, richly complex web of prior learning gives Kim, and other savants, our remarkable powers of recall.

Realizing that savants do not in fact possess photographic memories explains the limitations of savant memory. For example, Kim remembers trivia only from subjects in which he is interested; his memory is much less able to absorb a poem than a list of historical dates. In my own case, I have great difficulty remembering faces, even those of people I have known for many years. Consider for a moment the sheer complexity in every human face—not only the many individual details, but also the way in which a face and its features are never static but move continuously. For this reason, I remember the faces of friends and family by recalling recent photos of each of them.

Professor Simon Baron Cohen and his team at the Cambridge Autism Research Centre have looked more closely at the differences between my numerical memory and face recognition abilities. Compared to a range of nonautistic individuals around my age, my ability to remember and recall long strings of digits was highly developed, they found, whereas my ability to recognize photos of faces I had been shown only an hour before was significantly impaired. It is as though the part of the brain responsible for the warm, positive feelings of recognition that most people experience instantly when seeing a human face has been rewired in my head to feel the same emotions for numbers, instead.

My difficulty remembering faces illustrates the truth that,

despite what they might say or think, most people actually have very good memories considering, for example, the number of faces (and related personal statistics, gossip, and other trivia) that the average person can recognize and remember without difficulty over the course of a lifetime. The difference between the memories of savants and the general population is not so much how we memorize, but what. Numbers and facts are a lot easier for me to remember, and faces easier for most other people. A savant's ability to remember information is in many ways comparable to that of other people. Although it is much deeper and more elaborated, it is distinctly human all the same.

The Science of Memory

When scientists talk about memory, they use a range of words to describe the various forms of recall. Most researchers distinguish three major memory systems: *episodic memory* refers to recollections of particular episodes in one's life connected to specific times and places. *Semantic memory* describes our ability to remember more general information such as words, facts, and ideas, allowing a person to know, among thousands of other things, that 8 squared is 64, that Hinduism is a religion, and that your grandmother's name is Joan. *Procedural memory* involves our ability to learn skills and acquire habits: tying shoelaces, swimming, climbing a tree, and driving a car are all examples of this form of memory. Unlike the other two forms of memory, procedural memory does not normally require any conscious effort to recall.

Most everyday tasks require a seamless interaction of these memory systems. Learning to play a musical instrument, for example, requires episodic memory for the layout of the keys/strings or, when learning new melodies, semantic memory when recalling song lyrics, and procedural memory, which occurs after you have mastered the instrument and you play it without having to think about each note you play.

The psychologist Endel Tulving, one of the world's leading researchers on memory, argues that an intimate link exists between what he calls the rememberer and the remembered; he uses the analogy of "mental time travel" to describe the subjective experience of recalling events from our past. A growing number of scientific studies supports Tulving's view that our memories are not so much objective snapshots as they are subjective reformulations of our past experiences, shaped and influenced to a large extent by our particular thoughts and feelings at the time we form a memory.

For example, memory researchers recognize two distinct modes of remembering: *observer*, where the person recalls an event in which he sees himself, and *field*, where the person recalls an experience from the "outside," observing it from a perspective similar to the one had during the original event. The father of psychoanalysis, Sigmund Freud, wrote about the field-observer distinction and argued that observer memories were necessarily altered versions of the original events, because it is impossible to see ourselves in this way in reality. Researchers have discovered that we tend to observe ourselves in our memories of events that happened in the distant past, whereas we remember more recent memories from something like the original (field) perspective.

The cognitive psychologists Georgia Nigro and Ulric Neisser made the first scientific investigation of the field and observer modes of remembering in the early 1980s. In their study, they asked a group of subjects to remember events from their personal past while focusing on their feelings for each episode, and asked a second group to remember events while focusing on the objective circumstances of the episodes being remembered. They found that the subjects experienced more field memories when focusing on the events' objective circumstances, but more observer memories when they focused on their emotional reactions to the events. This finding suggests that the nature of our memories—whether we see ourselves as an actor in a remembered event, for example—is constructed at the point of recall, shaped by our goals and intentions as we attempt to remember a particular experience.

The subjective nature of memory is further demonstrated by the observation that different people remember very different aspects of the same objects or events. The French artist Sophie Calle was curious to discover what aspects of a painting remain in the memories of viewers who are familiar with it. In a 1991 experiment, she asked several members of staff at the Museum of Modern Art in New York to describe their recollections of René Magritte's painting, *The Menaced Assassin*, after it was removed from its usual place at the museum. A security guard recalled the painting as "a murder scene, men in dark suits, a pale woman, and dashes of red blood." Another, responsible for maintaining the painting, was able to describe it in much more detail: "It's a painting with a smooth surface . . . approximately five feet high and seven feet long. It is framed in a plain, dark, walnut-stained molding, something austere. I never liked it. I don't like stories in painting." A third, a curator, remembered the painting as follows: "It has a *film noir* sort of feel, a mystery novel to it . . . You have all those little clues that will probably lead you nowhere; there are men dressed in dark coats, and black bowler hats, the way Albert Finney was dressed in *Murder on the Orient Express*, placed in a room with a dead body. In the center, the one who seems to be the perpetrator is lifting the needle of a phonograph. Two weird-looking individuals are hiding to the side. There is a face looking from the balcony . . . And, when you look at her [the woman lying on the couch] carefully, you realize that the towel probably conceals a decapitated head."

The sheer range of memories related in Calle's project show that what we remember about an object like a painting depends on the thoughts and feelings we have about it at the time we view it. The particular aspects that we recall of an object or event depend also on the nature of the preexisting knowledge in our long-term memory: for example, the curator was able to remember a large amount of the painting's contents while referencing *films noir*, mystery novels, and the film *Murder on the Orient Express*, his familiarity with which likely helped him recognize and recall similar elements in the painting.

Remembering something from your past is not at all like pulling facts from your brain, but rather a reconstruction aided by your level of interest, knowledge, and emotion at the time of the event and the subsequent point of recollection. The neurologist Antonio Damasio contends that we remember in this way because there is no single area or location in the brain that contains a memory. Rather, different aspects of an experience activate different parts of the brain so that remembering involves a process of pulling these distributed pieces back together. Our memories are reformulations—rather than replications—of the original experience. Damasio also emphasizes that the very act of recollection affects the nature of our memories: "Whenever we recall a given object or experience we do not get an exact reproduction but an interpretation, a newly reconstructed version of the original."

Our memories thus are malleable rather than fixed, affected by the present as much as by the past. Indeed, each memory we have of our past is a new, emergent entity, created from the interaction of sensory fragments in our brains and the circumstances in which we remember the original event. Harvard psychologist Daniel Schacter has looked in detail at how the properties of our present environments can affect what we remember about the past. In one study, he asked college students to look at photos of people while hearing them speak in either a pleasant or irritating tone of voice. When later shown each photo again and asked to recall the person's tone of voice, students who saw a smiling face were more likely to remember the person as having spoken in a pleasant tone of voice, while photos of scowling faces tended to elicit memories of the person speaking in an unpleasant tone. This was in spite of the fact that there was no relationship between the people's facial expressions and their tone of voice. The students' memories were significantly influenced by the properties of the photos they were shown during recall.

Even our most intimate, personal memories are complex reconstructions rather than snapshots of our past. Martin Conway and David Rubin, researchers in autobiographical memory, suggest that autobiographical knowledge is multilayered, comprising three distinct levels that are arranged hierarchically. The highest level of this hierarchy contains memories spanning years or decades: our schooldays or living in a particular city, for example. The middle level contains memories of general events measured in days, weeks, or months: summer holidays or attending a wedding, for example. The bottom level contains memories of individual events that last seconds, minutes, or hours, such as the first time you ate a watermelon or a recent dream.

Scientific studies of these forms of memories reveal that each level serves a different function and may even be reconstructed by different systems in the brain. General event memories tend to be described more frequently than the other levels, probably because they are more easily recalled as they refer to repeated events in the past. For this reason, they form natural entry points into our personal reminiscences. Lifetime period memories—the highest level of the autobiographical memory hierarchy—help us to retrieve more specific autobiographical knowledge further down the hierarchy at the general or event-specific levels.

Supporting the thesis that memory is a complex, subjective reconstruction of the past, Conway and Rubin contend that no single representation stored in a person's brain has a one-to-one relationship with his autobiographical recollections. Instead, our memories are always constructed by combining bits of information from each of the three hierarchical levels of autobiographical knowledge.

In view of such evidence for the reconstructive complexity and emotional richness of our memories, a growing number of cognitive scientists are moving away from the well-known analogy of the brain as a computer. Nobel Laureate neurobiologist Gerald Edelman has articulated this viewpoint well, stating that human memory "involves a rich texture of previous knowledge that cannot be adequately represented by the impoverished language of computer science: storage, retrieval, input, output.

Remembering Better

The subjective nature of memory suggests that there may be as many ways to remember a piece of information as there are rememberers. Bearing this in mind, the following pages present a range of ideas that you can try for improving your own recall, based on well-supported scientific concepts and illustrated by my own personal experiences, as well as by those of several of my friends kind enough to share them with me for the writing of this section.

Perhaps the single most important part of improving your ability to remember effectively is to encode a learning experience deeply by attending to its meaning. Memory researchers have shown that the deeper a piece of information is registered in the brain, the more likely it will be successfully recalled later. For their studies, they use orienting tasks that guide encoding by requiring the subject to answer a specific question about the information to be memorized. Asking a question like, "Does *spider* have more vowels or consonants?" induces a shallow, nonsemantic encoding of the word where the subject does not have to consider its meaning, whereas asking "Is spider a type of animal?" prompts the subject to think about the meaning of the word "spider" and thereby constitutes a deeper, semantic encoding. Asked only the first question, subjects are much less likely to remember the target word "spider" on a subsequent recall test.

Interestingly, researchers have determined that only a certain kind of semantic encoding produces high levels of memory performance—an elaborative encoding that allows the person to integrate new information with preexisting knowledge. For example, asked the question, "Is spider a type of food?" the subject must pay attention to the meaning of the word in order to answer the question, yet does not integrate the target word with what he already knows about spiders. For this reason, the subject will show surprisingly poor recall in any subsequent test for whether the word "spider" was previously given to him.

Elaborative encoding is how experienced actors are able to memorize lengthy scripts with very high levels of accuracy. Rather than attempting to learn their lines by rote, the actors analyze scripts, questioning the underlying meaning of the material in order to better understand the motivations and goals of their characters. Studies confirm that when students are asked to use "all physical, mental, and emotional channels to communicate the meaning of material to another person, either actually present or imagined," their line retention improved significantly, compared to those who read the script for comprehension alone.

I can also learn and remember words easily, such as foreign language vocabulary, by using a form of elaborative encoding. Learning the French word "grenouille" (frog), for example, involved recognizing the ending "-ouille" as being the same as the one occurring in words I already knew, such as "la citrouille" (pumpkin) and "je chatouille" (I tickle), plus linking the beginning "gren-" with the English word "green" as frogs are often a green color. I give more examples of how I learn foreign language vocabulary in the next chapter.

Thinking of French words reminds me of the story related to me by a friend about his great-grandmother, who came to England from France after World War II to work as a cook for an English family. Not speaking any English, she used her own method to remember the words she heard spoken around the house, by linking them to what she knew best: the French names for different food items. In this way, she was able to remember a word like "good night" by relating it to "gousse d'ail" (garlic clove)!

Another friend of mine, who has an extremely good memory for the faces of the children with whom he went to school, shared his method for remembering another very important set of words: the names of people he meets. When he meets a new person and learns his name, he spontaneously thinks back to one of his schoolmates who had the same name; he is then able to make a connection between the visual image of his old friend from school and the person he has just met, which helps him to remember the new person's name.

Learning to use a comprehension strategy rather than rote memorization is a good way to remember lists of information. When I remember the names and dates of the kings and queens of England, or of the presidents of the United States, for example, I do so not by reciting them over and over to myself but by actively learning the circumstances behind the names and dates. This context brings understanding, which aids memory. Learning that Edward VI came after Henry VIII, and was followed by Mary I and then by Elizabeth I is made easier to remember when you know that males always succeed to the throne ahead of females, even when they are born later (as was the case with Edward). It is also useful to know that Edward was Henry's only son, which is why he was succeeded by women: Mary was the daughter of Henry's first wife, while Elizabeth the daughter of his second, which is why Mary succeeded Edward first. Knowing as well that presidents of the United States are elected to four-year terms and are allowed to serve a maximum of two terms (according to the Twenty-second Amendment to the U.S. Constitution), and that only Franklin Delano Roosevelt served more than two terms (elected four times between 1932 and 1944) is also useful when remembering the list of presidential periods in office.

As a child, I learned and remembered many things using my

imagination. Role-playing is a very effective way to encode new information, because it requires careful thought that derives from self-reflection: "How do I do this?" and "How would others do this?" are useful questions to ask yourself when learning something new. For example, one of the learning activities that I enjoyed playing with my siblings many years ago was to construct a government. As the "prime minister" I had to consult with allies, form policies, create a budget, collect taxes, organize elections, all based on what I had read in books about politics, economics, and voting systems. The act of playing out what I had learned helped me to effectively remember these new ideas and information.

Here is a further example of how using our imagination can help us to remember a sequence of facts, such as the list of planets in our solar system: Mercury, Venus, Earth, Mars, Jupiter, Saturn, Uranus, Neptune, and Pluto. Imagine that you are an astronaut leaving Earth to investigate the other planets that share our sun. As you leave Earth's orbit, you look back out of the window of your space rocket and see the sun and two planets between it and yourself: Mercury and Venus, small and red-hot because they are so close to the sun. Turning around and flying ahead, you arrive at the closest planet to Earth: Mars. Passing it, your view is obscured by the sheer mass of the middle planet of the nine: Jupiter; you have to change gears in order to avoid a collision. Negotiating your way around it, you marvel at the rings of its neighbors, Saturn and Uranus. As you reach the outer regions of our solar system you come to the final two planets, freezing-cold and greenish-blue because they are so far from the sun: Neptune and Pluto.

Music is another useful tool for aiding memory. Think back to the song you were likely taught in class to learn the sequence of letters that make up the alphabet, for example; the use of a simple tune makes it considerably easier to recall. In recent years, researchers have studied the relationship between music and the brain in an effort to understand why this should be.

When we listen to music, our brains perform an extraordinarily complex task—creating sound from the oscillations of air molecules that strike our eardrums; the molecules oscillate at specific rates, which the brain measures in order to construct an internal representation—a high or low pitch, based on that frequency. Our brains also derive meaning, even pleasure, from sound. Neuroscientist Daniel Levitin, director of the Laboratory for Music Perception, Cognition, and Expertise in Montreal, and his colleagues used a technique known as "functional and effective connectivity analysis" to study how music works in the brain. They discovered that when someone listens to a piece of music, the ears send signals not only to the auditory cortex—the region of the brain that processes the sound—but also to the cerebellum, one of the oldest parts of the brain (sometimes described as the "reptilian brain"), involved in coordinating the movement and timing of our bodies. Levitin's team found that, when a song begins, the cerebellum synchronizes itself to the beat and then attempts to predict where the beats will occur as the music continues, perking up when it guesses the right beat but even more so when the song violates the expectation in a surprising way. The cerebellum seems to find pleasure in performing these continuous tiny adjustments in order to remain synchronized.

The importance of the cerebellum to musicality and memory is especially intriguing to me, considering the special role rhythm has played in my life. As a young child my parents found that the only way they could soothe my frequent tantrums was to place me in a blanket and rock it rhythmically from side to side. Even to this day I will quite often rock gently backwards and forwards in my chair as I sit and read. I have loved listening to music for as long as I can remember—one of my earliest memories is of watching a music video for the group Dire Straits on the television in the early 1980s.

The link between music and memory makes sense for a number of reasons. As Levitin's research indicates, listening to music involves our brain making continuous predictions about the beat. In fact, our brains are continuously making predictions based on what they learn, and studies indicate that this predictive ability is key to having a good memory. Research carried out by John Gabrieli and his team at Massachusetts Institute of Technology showed that, while one specific brain region was very active when subjects were learning something, an entirely separate region lit up when they were predicting whether they would need to recall the information later on. Predicting is an important part of effective learning because it allows us to judge whether we have studied enough or need to review more. For this reason, people who make more accurate predictions are better learners.

Repetition is another essential ingredient in both music and memory. Repetition helps make a piece of music intelligible to the listener, even though we may not be consciously aware of the particular patterns within a song. Composers regularly invent new means of building repetition by varying the consistencies in their music, allowing it to assume many different forms and moods. The fact that repetition aids recall was demonstrated empirically by Hermann Ebbinghaus, a German psychologist who was the first researcher to carry out detailed experiments on human memory. In his classic 1885 book *Memory* he demonstrates that the retention of information improves as a function of the number of times the information has been studied.

Music also ties into memory systems that language alone does not. It powerfully articulates emotions, from happiness and joy to frustration and despair, and emotion plays a central role in the formation of strong, long-term memories. This is why in the centuries before the the printed word, knowledge was transmitted from generation to generation through epic songs rich in adventure, drama, and other activities associated with deeply emotional states.

A piece of music is most memorable when it is constructed to have a clear hierarchical structure, such as those found in traditional epic songs; in fact, such hierarchical organization aids memory storage and retrieval for any form of sequential information, from shopping lists to telephone numbers. To remember a series of digits such as 19897653113, for instance, we can group or "chunk" the numbers into higher-level subsets of information already familiar to us: 1989 (the year)—765 (descending order of numbers)—3113 (a numerical palindrome). The American psychologist George A. Miller published a famous scientific paper in 1956 entitled, "The Magical Number Seven, Plus or Minus Two," in which he demonstrated that short-term memory generally has a capacity of between five and nine "chunks" of information for a range of cognitive tasks. Converting the eleven-digit number example above into three distinct chunks helps us to get around the limits of our short-term memory capacity.

Any set of elements that can be associated with each other can become a chunk: musical notes that form a chord or letters that form a word, as examples. The chunking process represents an interaction and cooperation between short- and long-term memories; the association that helps form a memorable chunk is developed over time in long-term memory, yet its effect is to reduce the memory load in short-term memory by reducing the number of elements and "saving space."

Chunking can occur on multiple levels, meaning that a chunk can itself become an element in a larger chunk with whole chunks at lower levels in the hierarchy becoming part of chunks at higher levels (such as combining the chunk "765" with the concept of a numerical palindrome, as in "765567"). In this way the chunking process produces the kinds of structured hierarchies of associations that occur in memorable pieces of music.

Even though chunks are long-term memory structures, their size is determined by short-term memory capacity (7 ± 2 elements). Yet the potential amount of information that you can retain using hierarchical compression is considerable. With hierarchical rechunking we could theoretically remember an average of seven chunks with each chunk consisting of an average of seven elements (say, a sentence of seven words each consisting of seven letters). Assuming this process might yet extend to an average of seven levels of memory, it would create a hierarchy of 7 x 7 x 7 x 7 x 7 x 7 x 7 elements (e.g., a story of seven chapters, each consisting of seven sections, each made up of seven verses, each of which comprises seven paragraphs, each of these made up of seven sentences of seven words, each containing seven letters), with a total approaching a million individual items of information. In this context it is much easier to conceive of the possibility that a human mind might be capable of recalling over 22,500 consecutive digits of Pi, particularly when—as in my case—it is able to "chunk" groups of numbers spontaneously into meaningful visual images that constitute their own hierarchy of associations.

It is not surprising to scientists that my ability to visualize numbers as three-dimensional shapes (and experience words in colors) aids my memory for certain kinds of information. In 1969, the Canadian psychologist Allan Paivio demonstrated that subjects remembered concrete words such as "piano" much more easily than abstract words such as "justice," which are much harder to visualize. In another experiment, Paivio subsequently demonstrated that pictures were easier to memorize than words representing those pictures. And in his classic 1968 book, *The Mind of a Mnemonist*, the Russian psychologist Alexander Luria described in detail his decades-long study of a man called Shereshevsky who was able to visualize numbers and words and remember huge amounts of information, as I can. For example, Shereshevsky gave Luria the following description of how he sees the numbers 1 through 8:

[The number] one is a pointed number—which has nothing to do with the way it's written. It's because it's somehow firm and complete. Two is flatter, rectangular, whitish in color, sometimes almost a gray. Three is a pointed segment which rotates. Four is also square and dull; it looks like two but has more substance to it, it's thicker. Five is absolutely complete and takes the form of a cone or tower—something substantial. Six, the first number after five, has a whitish hue; eight somehow has a naïve quality, it's milky blue like lime.

Shereshevsky experienced images for words, too:

When I hear the word "green," a green flowerpot appears; with the word "red," I see a man in a red shirt coming toward me; as for "blue," this means an image of someone waving a small blue flag from a window.

Visual experiences of the kind shared by Shereshevsky and myself are the result of synesthesia—a cross-wiring of the senses in the brain that most often results in individuals being able to see letters of the alphabet and numbers in various colors. A 2002 study by researchers at the University of Waterloo in Ontario, Canada, provides further evidence that such synesthetic perceptions aid numerical memory. "C," a twenty-one-year-old female student and number-color synesthete, was presented with three 50-digit matrices of numbers and subsequently asked to recall the digits in each matrix. The first consisted of black digits; the second was composed of colored digits that did not match C's synesthetic perception of them, and the third contained colored digits that did match. C was able to accurately recall a high percentage of the numbers presented to her when the digits were displayed in black or in colors that matched her synesthetic perceptions of them, but was far poorer when the numbers were presented in colors that did not match her synesthetic perceptions.

This finding was replicated in 2004 by researchers Shai Azoulai and Ed Hubbard at San Diego's Center for Brain Studies, with an experiment that evaluated the relationship between my synesthetic perceptions of numbers and memory for them.

For me, the numbers 0 to 9 have different sizes ranging from 6 (the smallest) to 9 (the largest). The researchers presented me with two 100-digit number matrices (each for three minutes); one that presented the numbers in sizes that conformed to my synesthetic perception of them, and another that did not (for example, with large 6s and small 9s). Three days later I was able to remember 68 digits from the first but barely any from the second. I found the experience of being asked to read and later recall the numbers presented in the "wrong" sizes extremely dizzying and uncomfortable—rather like asking someone to read and recite in a language he does not know!

Accurate, reliable recall of information that you have learned depends not only on the elaborative encoding of it, but also on a subsequent process that scientists call consolidation. Researchers distinguish between two forms of consolidation: one that converts immediate or short-term memories that are seconds, minutes, or hours old into more enduring, long-term memories, and a second that operates over far longer periods of time: months, years, or even decades. For this reason some types of memory seem to become more resistant to disruption caused by brain injury or illness as the years pass. Accident survivors who have sustained serious head injuries, for example, exhibit this form of memory consolidation and typically retain memories from their distant past, but forget permanently the accident and the minutes preceding it and also lose temporarily memories of recent days, weeks, or months.

Recent research indicates that a good night's sleep is an important aid in the process of memory consolidation. In one study a group of subjects aged eighteen to thirty were divided into two groups and given twenty word pairs to learn. The group given the words to learn at 9 p.m. and retested the following morning, after a night's sleep, performed better than those who learned them at 9 a.m. and were tested on them at 9 p.m. the same day.

On the basis of such studies, researchers have identified three stages that make up a memory's "life cycle." The first, stabilization, takes around six hours during which time the memory is particularly vulnerable to being lost. The second, consolidation, occurs during sleep. The third and final stage, the recall phase, is when the memory is once again available, accessible, and even modified.

Gaining access to our memories once they have been consolidated often depends on retrieval cues (something that helps to "jog" a person's memory). Scientists have found that it is not the literal similarity between the conditions in which a memory was formed and is subsequently remembered that matters most, but rather whether a cue reinstates a person's subjective perception of an event including whatever thoughts or inferences occurred at the time the memory was being formed. For example, researchers who gave subjects the rather gruesome sentence, "the weapon was protruding from the corpse," found that many of the subjects inferred that the weapon in question was a knife, and that later giving them the word "knife" helped them to recall the sentence more often than when they were given the word "weapon," even though the latter appeared in the original sentence.

Retrieval cues are often dependent on the context of the memory. For example, you hear a song somewhere and remember something completed unrelated to it (because the song acts as a cue to the original context in which you heard it). Researchers have found that you are more likely to remember something if the place or situation in which you are trying to recall the information bears some resemblance—color or smell, for example—to where you originally learned it. Moving students from one classroom, where they have learned their study material, to another for examinations is likely to have a negative impact on their ability to recall what they have learned.

The other kind of retrieval cue is state-dependent, where re-creating the original state of mind you experienced during the learning period can help retrieve information during a subsequent attempt at recall. In a 1977 experiment volunteers were given an alcoholic or nonalcoholic drink before studying a list of words. The following day they were asked to recall as many of the words as they could while either drunk or sober. Those subjects who consumed alcohol during both the learning and recall stages remembered more than those intoxicated during only the study phase.

Highly associative learning, such as Kim Peek's complex interwoven "web" of facts and figures or my own hierarchy of numerical and word associations, is naturally cue-rich, which helps our recall. Two simple examples from my own experience: three years ago I visited a Welsh-speaking part of Wales where the road signs were all marked with words in the Welsh language. In a subsequent attempt more than a year later to recall one of those words in a conversation with a friend, I was able to visualize the shape and picture of the sign, which in turn helped me to retrieve the "shape" of the word itself: medium-length with the letter "o" in the center. After several moments the word appeared in my mind: "henoed" (meaning "elderly people").

The second example occurred during a quiz event in which I participated with a group of friends. One of the questions asked was: "Who wrote the novel *The Prime of Miss Jean Brodie*?" Although I had never read the book, I knew it was famous and that I had likely read the author's name somewhere at some time. After a few seconds my memory had retrieved the cue name Starsky (from the 1970s TV detective show *Starsky and Hutch*) and after a few more seconds I was able to move from that to the similarly shaped name "Spark" and then to the answer: Muriel Spark.

Déjà Vu and Other Distortions

Sometimes the retrieval cues our brains use to remember past experiences can distort our memories. In the experience of déjà vu (French for "seen before"), the individual has the strange sensation of having already been in the present situation at some time in the past, even though he knows he has not. One possible explanation for the déjà vu phenomenon is that aspects of the current situation act as retrieval cues for a memory of a similar, previous experience, thereby causing an eerie sense of familiarity.

Scientists have studied the nature and causes of memory distortion for decades. The British psychologist Sir Frederic Bartlett was the first to demonstrate conclusively that a person's memory for a complex event is shaped, and sometimes distorted, by his personal beliefs, feelings, and inferences. In his most famous experiment, published in 1932, Bartlett asked a group of subjects to read a story entitled, "The War of the Ghosts," based on a tribal legend of the Chinook people in the Columbia River area of North America. The subjects were subsequently asked to write down their recollection of the story on a number of different occasions after increasingly long periods of time. The narratives that the subjects wrote were significantly shorter than the original story and became further abbreviated with each reproduction. Unfamiliar names of geographical locations were lost and unconventional phrasing within the story was modified into more conventional language. Bartlett's subjects changed the story to fit their existing knowledge and expectations, and subsequently remembered these revised versions of the original.

More recent studies confirm Bartlett's theory that our memories continuously and subconsciously interact with our brain's "bank" of preexisting general knowledge built up from past experiences, and that memory distortion can occur when inferences based on this past knowledge creep into our attempts to remember something. For example, spend two minutes studying the following list of words: candy, sour, sugar, bitter, good, taste, tooth, nice, honey, soda, chocolate, heart, cake, eat, and pie. Now turn away from the page and write down as many of the words as you can remember.

Without looking back at the previous paragraph, consider the following three words in italics and try to recall whether they appeared on the list that you just learned: *taste, point, sweet.* When psychologists Henry L. Roediger and Kathleen McDermott of Washington University asked a group of subjects whether the word "sweet" appeared on the list many said that it did—even though it did not. Not only did these subjects believe that the word "sweet" was on the list but they also claimed to remember it vividly.

Why do these subjects remember the word "sweet" being on the list? Being presented with so many words associated with the word "sweet" possibly activates the general category of "sweet things" in your mind. When Kim Peek was given the same test by Professor V. S. Ramachandran, of San Diego's Center for Brain Studies, he remembered correctly that the word "sweet" was not on the original list. This is probably because savants' more detail-oriented perception and memory are less susceptible to subconscious inferences based on general categories.

The nature of memory distortion is especially important in the courtroom. In many cases, an eyewitness's testimony is the main evidence by which a jury decides a suspect's guilt or innocence. Elizabeth Loftus, an American psychologist, conducted a study that demonstrates that eyewitnesses can reconstruct their memories depending on the questions they are asked in order to retrieve them. Subjects were asked to watch a videotape of a car accident involving two vehicles and later given a questionnaire about the incident. One of the questions: "About how fast were the cars going when they hit each other?" was altered for different groups of subjects so that the verb "hit" was replaced with "smashed," "collided," "bumped," or "contacted." Even though all the subjects watched the same videotape, their speed estimates varied considerably, depending on how the question was asked. The average speed estimate was 32 mph when the verb was "contacted," 34 mph when it was "hit," 38 mph when it was "bumped," 39 mph when it was "collided," and 41 mph when it was "smashed." In a follow-up study one week later, the subjects were asked whether there had been any broken glass at the accident scene (there had not been). Those given the question using the verb "smashed" were more than twice as likely to "remember" broken glass at the scene than those given the question with the verb "hit."

Perhaps the most striking example of memory distortion occurs in the phenomenon of false memory, where a person remembers events from his life—sometimes in vivid detail and with considerable emotion—which he never experienced. A remarkable case of false memory involving a woman calling herself "Anna Anderson" is particularly familiar to me because I researched and wrote a lengthy essay on it for my high school history exam. Franziska Schanzkowska, Anna Anderson's real name, was born in Pomerania in modern-day Poland in 1896 and worked in a Berlin munitions factory during the First World War. Following an incident at the factory in which Schanzkowska accidentally dropped a grenade that killed one of her coworkers before her eyes, she suffered from shock and was sent to a sanatorium. In February 1920, a short time after her release, Schanzkowska threw herself from a canal bridge and was subsequently rescued by a passing police officer and taken to a mental hospital in Dalldorf. She was nicknamed "Fräulein Unbekannt" (Miss Unknown) by the nurses there because she rarely spoke and refused to give any information about herself.

Schanzkowska spent two years at the hospital, much of the time passed reading articles in popular magazines about the Russian royal family, who had only a short time before been executed (during the summer of 1918) on the orders of Lenin's new Bolshevik government. A fellow psychiatric patient claimed that she recognized Schanzkowska as one of the grand duchesses, based upon one of the magazine photos, and shortly afterwards Schanzkowska began calling herself "Anastasia," believing that she was indeed one of the tsar's daughters. She claimed she had been rescued from the firing squad by one of its members sympathetic towards her and that they had traveled together in a cart across Europe to escape the Bolshevik soldiers.

Subsequently taken in by various members of Berlin's sizable Russian émigré community, Schanzkowska received numerous visits from those who had personally known the real Grand Duchess Anastasia, all no doubt hoping that she had indeed escaped alive. During these meetings she was shown many photos and items that had once belonged to the Russian royal family and many anecdotes from Anastasia's childhood were also shared with her, in the hope of some flicker of recognition. In fact, these meetings formed much of the material for Schanzkowska's false memories of being Anastasia, though much of this information and her subsequent "memories" based on them were inaccurate. Olga Alexandrovna, an aunt of the real Anastasia, met Schanzkowska in the mid-1920s and later shared details of the experience with her biographer:

> She [Schanzkowska] had a scar on one of her fingers and she kept telling everybody that it had been crushed because of a footman shutting the door of a landau too quickly. And at once I remembered the incident. It was Marie, her elder sister, who got her hand hurt rather badly, and it did not happen in a carriage but on board the imperial train. Obviously someone, having heard something of the incident, had passed a garbled version of it to her.

In 1938 Schanzkowska's lawyers filed a suit in the German courts, seeking to establish her legal identity as the Grand Duchess Anastasia. The judges heard from numerous handwriting experts, forensic scientists, and historians who had scrutinized vast quantities of photographs and documents. The case dragged on until 1970 when the court finally ruled that Schanzkowska's claim was *non liquet* ("not proven").

In the late 1960s Schanzkowska, who by then had assumed the name "Anna Anderson," moved to the United States and married one of her supporters, an American history professor. She continued to share her "memories" of her childhood as Anastasia with journalists until her death in 1984. A decade later, DNA tests performed on a sample of Schanzkowska's preserved tissue matched a blood sample given by Schanzkowska's great-nephew, finally settling the issue for good and proving conclusively that she could not have been the tragic Russian princess.

Forgetting

"Anna Anderson" was able to forget her unhappy past as a Polish factory worker and live most of her adult life under a new, glamorous identity fueled by her "memories" of a childhood spent in Russia's royal palaces, but for most people forgetting past experiences can be frustrating, bewildering, and even traumatic.

The nature of forgetting is one of the most studied subjects in experimental psychology. Hermann Ebbinghaus, the nineteenth-century pioneer in memory research, was the first to develop precise methods to measure how we forget. Using himself as his sole subject, he created dozens of lists of nonsense syllables which he then learned by repeating them over and over to himself until he could recite them back without error. He then tested his memory over various periods of time ranging from twenty minutes to a month, measuring how much he had forgotten at each time interval. By conducting his experiment with many lists, he discovered that the rate of forgetting was relatively consistent and exponential: occurring rapidly at first before leveling off over time. Psychologists have since been able to graph the average rate of forgetting as a curve for a wide range of types of information.

The things people forget can be surprising once a sufficient period of time has passed. For example, in a 1993 study, the psychologists JoNell Usher and Ulric Neisser interviewed a group of subjects whose families confirmed all shared a sad fact in common: the death of a family member when they were four years old. Yet, when asked about the incident, over 20 percent of the subjects failed to recall a single detail of it.

Why do we forget? The oldest theory is that forgetting is caused by a disintegration of memory traces formed in the brain as we learn information. This theory is no longer held by most psychologists for two main reasons. First, the "decay theory of forgetting" only describes the fact of forgetting; it does not explain the processes behind it. Second, the fact that a person can forget information at one point only to retrieve it perfectly well at a later one could not happen if all memories inevitably deteriorated over time.

Another suggested cause of forgetting is "repression," an idea introduced in the late nineteenth century by Sigmund Freud. According to his theory, people push away unpleasant experiences into their unconscious minds, where the "repressed" memories can continue to unconsciously influence a person's behavior and result in unpleasant side effects, from slips of the tongue to sustained illness. Most evidence for Freud's theory is found in case studies, however, which are open to many different interpretations. Many memory researchers are skeptical that repression can explain why people forget.

Psychologists' most popular, current explanation of forgetting is that it is caused by interference, when information we want to remember gets confused or disrupted by other information in our long-term memory. There are two forms of interference: "proactive interference" occurs when previous learning or experience interferes with your ability to learn new information. An example would be studying Spanish in one school year and Italian in the next. Given a word test for Italian later on, it is likely that your prior learning of Spanish might interfere with your ability to remember the correct Italian translations. "Retroactive interference" occurs when newly learned information interferes with your ability to remember earlier information or experiences. For example, try to recall what you ate for supper five days ago. The suppers you have eaten in the intervening days probably interfere with your ability to remember this particular event. Both types of interference can significantly impair your ability to remember.

Not all forgetting is bad, however. After all, no one needs to remember everything that has ever happened to him. The American psychologist and philosopher William James agreed, writing that: "In the practical use of our intellect, forgetting is as important a function as recollecting." Others have argued that forgetting helps prevent obsolete information, such as old telephone numbers, from interfering with the recall of current, relevant information.

What if we could in fact remember every detail of our past without forgetting a single thing? Jorge Luis Borges, the Argentine writer, wrote a fantasy short story about just such a person entitled, "Funes the Memorious." Borges's story centers around the character of Ireneo Funes, an Uruguayan teenager who finds that following a horse-riding accident, he is able to perceive everything in full detail and remember it all. For example, the story's narrator gives Funes several Latin books and days later finds that he has learned them perfectly, reciting from memory "the first paragraph of the twenty-fourth chapter of the seventh book of the *Historia Naturalis* (by Pliny the Elder)" as the narrator enters his room. Funes explains that he perceives everything around him in acute, chaotic detail:

He remembered the shapes of the clouds in the south at dawn on the 30th of April, 1882, and he could compare them in his recollection with the marbled grain in the design of a leather-bound book which he had seen only once, and with the lines in the spray which an oar raised in the Rio Negro on the eve of the battle of the Quebracho.

Funes finds his inability to forget intolerable as his mind clogs with thousands of unimportant details, preventing him from generalizing or thinking for himself. He finds it difficult even to sleep at night because he remembers: "every crevice and every moulding of the various houses which [surround] him."

Though forgetting is a natural, beneficial part of our mind's processes, a sudden loss of memory—as can occur following a head injury or illness—can prove devastating to the individual and those around him. Amnesia is a memory disorder that results in an abnormal degree of forgetfulness and/or an inability to remember past events, which can result from a range of causes, including brain damage, stroke, trauma, or disease. Recent reseach indicates that amnesiacs are equally unable to imagine the future, because we use memories from past experiences to help form imagined future scenarios in our minds.

Unable to recall their past or project themselves into the future, amnesia sufferers are continuously caught in the present. A dramatic example of this can be seen in the tragic case of Clive Wearing, a British musicologist, who suffered acute and permanent memory loss following a brain infection in his forties. For the past twenty years since his illness, Wearing has forgotten everything that just happened. Unable to remember for longer than a few minutes at a time, he spends his days continuously believing he has just awakened from a deep sleep. Wearing's case exemplifies the complexity of memory: he can recognize his wife, but has no recollection of their wedding day. He recognizes his children, too, but cannot remember how many he has. A music producer and choirmaster in his former life, he can still play the piano but has no memory of old songs and so is unable to improvise or create anything new.

Memory loss that results from various forms of dementia can occur more gradually, leaving sufferers with an ever-dimming awareness that they are slowly losing the memories that make up the stories of their lives, and even their sense of self. British novelist Iris Murdoch descended into Alzheimer's disease in her final years, at first believing the symptoms were writer's block. She described having the condition as like "being in a very, very bad quiet place, a dark place."

Many people fear a general decline in memory as they get older, but although aging can indeed impair memory, older adults' ability to remember well varies widely across different situations. For example, when older people are shown two separate lists of sentences and later asked to recall whether a sentence appeared on the first list or the second, they perform less well than young people. Yet when the sentences are displayed on the left or right side of a screen older people remember where the sentences appeared as well as young people.

Aging has a disproportionately severe effect on the frontal regions of the brain, with shrinkage and reduction of blood flow more pronounced in the frontal lobes than other brain regions. Areas of this part of the brain play a critical role in certain forms of remembering, which may explain why aging can result in a performance decline for some memory tasks but not others. For example, memory tasks that rely on source memory—the ability to remember from where a particular piece of information was learned—have been shown to be far trickier for older adults than for young people. In one study, a series of fictitious facts was given to old and young people by either a man or a woman. The researchers found that the elderly had much more difficulty remembering whether the man or the woman had given them the information, even though they could recall the facts themselves correctly.

Fortunately, when older adults use their extensive preexisting knowledge to imaginatively learn new information, and are subsequently given cues to help them retrieve these memories, they can remember about as well as young people. This is because our memory for facts and associations is generally well preserved as we age.

Reminiscing is an especially common form of memory activity among the elderly, demonstrating that our ability to recollect and engage with our past can remain largely intact throughout our lifetime, yet it is often considered negatively as "living in the past." Research by gerontologists (scientists who study how and why we age) refutes this view, indicating instead that reminiscence is a normal, healthy part of the aging process. Studies show that older adults who tend to reminisce (specifically, reminiscences that focus on previous plans or goals or on reconciling the past with the present) are less prone to depression and are mentally healthier than those who do not.

Perhaps because of this love for reminiscing (and the benefits that accrue from it), older people often make impressive storytellers. In many societies, elderly adults are encouraged to pass on their stories—and the knowledge and ideas they contain—to the younger generation. For example, elders in Native American tribes were greatly respected as sources of cultural memories, often told in the form of creation stories that passed on vital information such as the tribe's origins, how to hunt, treat the environment with respect, and so on. Those arrogant enough to ignore the elders' stories were said to be doomed to repeat the mistakes of the past. Unfortunately, the arrival of missionaries and Western culture did much to destroy the Native Americans' memory-centered traditions.

Indeed, the role and significance accorded to memory in many Western countries has fallen sharply, as computers and other gadgets are seen as replacing the need to commit experiences or information to heart. But computers make poor substitutes for the sophisticated, contextual, and interpersonal nature of our memories. To remember is human, because the past—personal and collective—is the wellspring of our present, and our future.

4

A World of Words

I am a linguaphile—a lover of words and language. *Love* is the right word for how I feel about language—I learn and speak numerous languages as much for their beauty as for their utility. Certain words or combinations of words are especially beautiful and affecting to me, including: "buttercup," "ljósmóðir" ("midwife" in Icelandic, literally: "light mother"), and the Finnish "aja hiljaa sillalla" ("drive carefully on the bridge").

My foreign language learning did not begin until I was nearly in high school, coming as I do, like so many Englishmen, from a stubbornly monolingual family. With few native speakers to help me, much of my early learning was limited to books and cassettes. At first, my sole motivation was my fascination with the grammatical patterns and beautiful words I found in each language I studied. For example, one especially fond childhood memory is the following nursery rhyme (a version of "Incy Wincy Spider") being sung to me by my Finnish neighbor, which I remember to this day:

> Hämä-hämähäkki kiipes langalle.
> Tuli sade rankka, hämähäkin vei.
> Aurinko armas kuivas satehen.
> Hämä-hämähäkki kiipes uudelleen.

Knowing many languages has given me experiences and opportunities I could never have had otherwise. For example, I learned Lithuanian at nineteen during a yearlong stint in the country as a volunteer English teacher. I loved learning how to use certain diminuitive suffixes (such as "-let" in the English "booklet") in Lithuanian to express affection for close friends: Birutė (my best friend) was to me "Birutėlė" (meaning something like "little or dear Birutė"). My Lithuanian friends, in turn, called me "Danieliukas" ("little Daniel"). As a speaker of Esperanto (a universal language created in the nineteenth century to ease communication between speakers of other languages), I can understand and appreciate the humor in words such as "bonantagulo" (a lazy Esperanto learner), "krokodili" (to speak in a foreign language when Esperanto would be a more appropriate medium; literally "to crocodile"), and "volapukajho" (nonsense, from the name of an artificial language that preceded Esperanto).

Two further brief examples of the pleasure that language learning has given me over the years include reading the children's classic *Le Petit Prince* in the original French, and having a short conversation in rudimentary Gaelic with one of the language's few remaining native speakers during a holiday on the Isle of Skye in Scotland. The point I want to make here is that words and grammars are not cut off from life's rhythms. Rather, they are inextricably bound up in all our diverse, day-to-day experiences. Languages have helped me to learn more and more about what it is to be human—about the human condition in general and about distinct local perceptions, subtle insights, and emotional sensibilities.

The sheer depth and complexity of language is a further source of fascination. Think, for example, of the possible number of words in the English language (often cited as the wordiest of the world's languages). Richard Sproat, a computational linguist, compiled all the distinct words used in the tens of millions of words of text from Associated Press news stories for 1988.

His list totaled some 300,000 distinct word forms, among them gems such as "armhole," "boulderlike," and "traumatological."

In fact, the number of possible words in a language is infinite, due to the combinatorial power of a language's word-building rules: think of the unending number of words that you can generate simply by continuously reaffixing "great"to "grandfather": great-grandfather, great-great grandfather, great-great-great grandfather, and so on. Another example: take the noun "ration" (which historically meant "reasoning") to which we can add the suffix "-al" to create the adjective "rational." Next we add "-ize" to create a verb: "rationalize." We get back to a noun by using the suffix "-tion": "rationalization." In principle, there is no reason why we cannot go on creating new words in this way indefinitely, continuing with "rationalizational" (for which the search engine Google finds no fewer than 127 results).

Sentences are also potentially infinite in number. This is because a language's grammar is an example of a "discrete combinatorial system," in which the discrete elements (words) can be used, combined, and rearranged to create distinct, larger structures (sentences). An example of a potentially infinite sentence comes from a popular British children's rhyme (I will restrain myself to the opening few lines):

> This is the house that Jack built.
> This is the malt that lay in the house that Jack built.
> This is the rat that ate the malt
> That lay in the house that Jack built.
> This is the cat that killed the rat
> That ate the malt that lay in the house that Jack built.
> This is the dog that worried the cat
> That killed the rat that ate the malt
> That lay in the house that Jack built.

The marvel of language, with its potentially infinite number of words and sentences, is that anyone can speak it at all. Yet speak it we do, with few exceptions, and regardless of family background, level of education, culture, class, or race. In fact, the human capacity to acquire and use language is a profoundly outstanding intellectual achievement we all share.

To demonstrate further how remarkable this achievement is to us, we need only look at the number of words an average person knows. When asked, most people guess a figure in the hundreds or low thousands, but this proves to be a gross underestimation. Psychologists estimate the actual figure by randomly sampling words from a dictionary, which they present to each testee with a set of potential synonyms to choose between. Adjusting for guesses, the psychologists multiply the proportion of the sample correctly recognized by the size of the dictionary to arrive at an estimate of that person's vocabulary size.

Using this technique, the psychologists William Nagy and Richard Anderson calculated that the average American high school graduate knows 45,000 words—three times as many as appear in the collected plays and sonnets of William Shakespeare. For those who read frequently, the figure is likely to be a lot higher still.

Consider then how quickly these words must have been learned to reach a vocabulary of this size by age eighteen. With word learning beginning around age one, the high school graduates must have been learning an average of seven to eight words every day, continuously, for seventeen years. Harvard neuroscientist Steven Pinker sums up this finding by describing children as being "lexical vacuum cleaners."

The Language Instinct

Children do not wake up one morning with a fully formed grammar and vocabulary in their heads; rather, they acquire language in stages, with each stage approaching more and more closely the adult language. Scientific studies of how children acquire their first language in different parts of the world suggest that these learning stages are very similar and possibly universal.

The first stage in an infant's development of language begins around six months, when he begins to babble. Deaf as well as hearing children babble, indicating that it does not depend on auditory input from parents. Scientists are not sure what role babbling plays in the child's subsequent acquisition of language, but one theory is that it is during this period that the child learns to distinguish between the sounds that are part of his language and those that are not.

Next comes the "holophrastic" (expressing the ideas of a phrase or sentence in a single word) stage when, sometime after the child's first birthday, he begins to use the same combinations of sounds repeatedly to refer to the same thing. The child has learned that sounds are related to meanings and is producing his first words. These are generally naming words, such as "dog," "car," and "ma." During this stage, the infant often uses words in ways that are either too narrow or too broad (known as "underextensions" and "overextensions"): "bottle" used only for plastic bottles; "teddy" used only for a particular bear; "dog" used for cats, cows, and lambs, as well as dogs. Such usages develop and change over time from child to child.

Around his second birthday the child begins to produce two-word sentences. Here are some examples of these kinds of sentences:

"Hi, Mommy."
"Bye-bye, boat."
"Clock mantelpiece."
"Katherine sock."
"It ball."

There is little grammar produced during this "two-word" stage. Pronouns (he, we, they), for example, are rarely heard, although many children use "me" to refer to themselves.

The final stage leading up to adultlike speech is known as the "telegraphic stage," where the child starts to produce sentences of two, three, four, five words, or longer. The child's first full-length sentences usually lack "function" words such as "the," "to," and "is"— only "content" words are used. For this reason, children at this stage sound as though they are reading a telegram (hence the term "telegraphic speech" used to describe it). Examples include:

"Kathryn no like celery."
"Want lady get chocolate."
"Cat stand up table."
"He play little tune."
"No sit there."

These sentences are not strings of randomly distributed words but show similar structures to those produced by adults, revealing the child's early grasp of the principles of sentence formation.

As a child's language develops, he will gradually acquire more and more of the grammatical features of adult language. The ending "-ing"—used to indicate a continuous or ongoing action—is among the earliest found in young children's speech, as in the sentence: "Me going park." Prepositions, words such as "in" and "on," next appear, followed by the regular plural ending "s," e.g., "book—books." By age five, most children are able to talk with confidence and express themselves fluently or near fluently.

Various theories attempt to explain how children acquire adult language. One is that the child imitates what he hears. Though some imitation does indeed occur during the learning stages, the sentences produced by young children show that they are not imitating adult speech, for adults would never put together sentences such as, "A my pencil" or "two foot."

Another reason to discount the idea that children acquire

their language solely through imitation is the typical resistance of a child's grammar to correction from an adult:

> Child: "My teacher holded the baby rabbits and we patted them."
> Adult: "Did you say your teacher held the baby rabbits?"
> Child: "Yes."
> Adult: "What did you say she did?"
> Child: "She holded the baby rabbits and we patted them."
> Adult: "Did you say she held them tightly?"
> Child: "No, she holded them loosely."

An equally unlikely theory suggests that children learn to produce correct sentences by positive reinforcement when they say something right, and negative reinforcement when they say something wrong. Yet children do not correct their sentences, even when their mistakes are explicitly pointed out to them:

> Child: "Nobody don't like it."
> Mother: "No, say 'Nobody likes it.'"
> Child: "Nobody don't like it."
> Mother: "Now, listen carefully. Say 'Nobody likes it.'"
> Child: "Oh, nobody don't likes it."

In fact, young children seem to acquire their language by finding patterns in what they hear, and then use these rules whenever they can. A side effect of this rule-based approach to learning is that the child sometimes "overgeneralizes," for example, treating irregular verbs and nouns as though they were regular: "goed" (instead of "went"), "singed" (instead of "sang"), "mouses" and "sheeps." Only subsequently does he learn that there are exceptions to the rules, refining his generalizations and revising his knowledge of the rules accordingly.

Because children develop complex grammars both rapidly and spontaneously, scientists have long believed that they must be innately equipped with a special ability to acquire language.

Charles Darwin, the father of evolutionary theory, agreed with this view, describing language as being as instinctive in humans as the upright posture. A century later, the American linguist Noam Chomsky further developed and popularized this idea, arguing that the human brain contains a set of unconscious constraints (which he called "universal grammar") for organizing language. He explained his thinking as follows:

> It is a curious fact about the intellectual history of the past few centuries that physical and mental development have been approached in quite different ways. No one would take seriously the proposal that the human organism learns through experience to have arms rather than wings, or that the basic structure of particular organs results from accidental experience . . . Human cognitive structures, when seriously investigated, prove to be no less marvelous and intricate than the physical structures that develop in the life of the organism. Why, then, should we not study the acquisition of a cognitive structure such as language more or less as we study some complex bodily organ?

According to Chomsky's theory of universal grammar, the mental process—in any language—by which a sentence is perceived as grammatically correct while another is not is universal and independent of meaning. Thus, native English speakers can perceive immediately that a sentence like "I milkshake want a" is not correct English. At the same time we recognize that a sentence like "colorless ideas sleep furiously" is grammatically correct English, even though it does not make any sense.

Until the 1960s, when Chomsky's theory was introduced, language scientists generally held that children came into the world with minds like blank slates, and that they learned language solely by simple interaction with their environment. In challenging the "blank slate" view of the human mind, Chomsky and his supporters revolutionized our modern understanding of the nature and process of language acquisition.

One of the strongest pieces of evidence in support of Chomsky's ideas is the emergence of Creole languages in various parts of the world. We can trace the historical roots of many of these back to the period of the slave trade, when slaves and laborers from numerous countries and language backgrounds were mixed together on plantations. In order to communicate with one another, these workers had to develop a simplified language known as "pidgin," a blend of words from the different languages used on the plantation. These pidgins were extremely simple, lacking many of the grammatical features—such as consistent word order, affixes, or tenses—that normally occur in a native speaker's language. Often the slaves' children were isolated from their parents and raised by workers who spoke to them in the pidgin. Not content simply to imitate the fragmentary strings of words they heard, the children spontaneously introduced grammatical complexity into their speech, creating a brand-new, expressive language (known as a Creole) in the space of a single generation.

The linguist Derek Bickerton has hypothesized that Creoles, being largely the product of the children's minds, should provide a particularly clear insight into the brain's innate grammatical machinery. He argues that Creoles from around the world share remarkable similarities, and perhaps the same underlying grammar, even though they evolved from different source languages. Bickerton suggests that this basic grammar even shows up in the sentences produced by young children of more embellished languages, such as when English children say things like: "Why he is leaving?" and "Nobody don't likes me," which are grammatical in many of the world's Creole languages.

If language is indeed something that emerges spontaneously from the human mind, rather than merely a response to our environment, it should have an identifiable seat in the brain.

Although the search is still ongoing for a "language organ" or "grammar gene," there are examples of brain conditions that affect language while sparing overall cognition, such as Broca's aphasia, a syndrome caused by damage in the lower part of the left frontal lobe. The condition is characterized by slow, deliberate speech that lacks many features of normal grammatical sentences. The neuropsychologist Howard Gardner interviewed David Ford, a thirty-nine-year-old Coast Guard operator who had suffered a stroke three months previously and subsequently developed Broca's aphasia:

> I asked Mr. Ford about his work before entering the hospital. "I'm a sig . . . no . . . man . . . uh, well, . . . again." These words were emitted slowly, and with great effort. The sounds were not clearly articulated; each syllable was uttered harshly, explosively, in a throaty voice. With practice, it was possible to understand him, but at first I encountered considerable difficulty in this. "Let me help you," I interjected. "You were a signal . . ." "a signal man . . . right," Ford completed my phrase triumphantly. "Were you in the Coast Guard?" "No, er, yes, yes . . . ship . . . Massachu . . . chusetts . . . Coastguard . . . years." He raised his hands twice, indicating the number "nineteen." "Could you tell me, Mr. Ford, what you've been doing in the hospital?" "Yes, sure. Me go, er, uh, P.T., nine o'cot, speech . . . two times . . . read . . . wr . . . ripe, er, rike, er, write . . . practice . . . getting better." "And have you been going home on weekends?" "Why, yes . . . Thursday, er, er, er, no, er, Friday . . . Bar-ba-ra . . . wife . . . and, oh, car . . . drive . . . purnpike . . . you know . . . rest and . . . tee-vee." "Are you able to understand everything on television?" "Oh, yes, yes . . . well . . . almost." Ford grinned a bit.

Aphasics have a style of speech in which content words (nouns, verbs, and adjectives) are used, but many function words (such as "the," "and," "or") are left out. Broca's aphasia can also produce some particularly unusual forms of agrammatism (the inability to produce grammatically correct sentences). For example, Ford could read and use the words "bee" and "oar" but had much more difficulty with the more common "be" and "or."

Aside from his linguistic handicap, Ford's cognitive abilities were unaffected, as Gardner notes in his report on the case: "He was alert, attentive, and fully aware of where he was and why he was there. Intellectual functions not closely tied to language, such as knowledge of right and left, ability to draw . . . to calculate, read maps, set clocks, make constructions, or carry out commands, were all preserved."

The opposite to Broca's aphasia, where language impairment occurs alongside preserved cognitive function, can be seen in a condition known as "chatterbox syndrome" in which the individual suffers from a form of mental retardation, yet possesses relatively impressive language abilities. A famous case study involved an American teenage girl called Laura, who was unable to look after herself or even know her age: "I was sixteen last year, and now I'm nineteen this year," she once announced. Yet Laura was able to speak fluently in grammatically correct (if nonsensical) sentences. Here is a passage to give an idea of Laura's language skills:

> "It was kind of stupid for Dad, an' my mom got um three notes, one was a pants store, [of] this really good friend, an' it was kind of hard. An' the police pulled my mother out of [there] an' told the truth. I said, 'I got two friends in there!' The police pulled my mother [and so I said] he would never remember them as long as we live! An' that was it! My mother was so mad!"

Laura is not just parroting sentences that she hears other people using, as the following examples of her speech demonstrate:

"It was gaven by a friend."
"These are two classes I've token."
"I don't know how I catched it."

These errors show that Laura is creating her sentences by herself, in spite of her mental retardation. Her case provides further evidence for Chomsky's argument that natural processes within the human brain shape our ability to create and use language.

Language Universals

If language really is structured according to an unconscious universal grammar within the human brain, then various fundamental characteristics ought to be shared by most if not all of the world's languages. Many linguists have been researching the "universal tendencies" among languages over several decades and have compiled an impressive body of evidence from thousands of languages around the globe.

The most famous and influential of these researchers was the American linguist Joseph Greenberg, author of a 1963 research paper in which he cited forty-five language universals from an investigation of thirty languages across five continents including: Basque, Welsh, Yoruba, Swahili, Berber, Hebrew, Hindi, Japanese, Maori, Quechua (a descendent of the language of the Incas), and Guarani (an American Indian language). Among the universal tendencies he discovered were:

The subject of a sentence almost always precedes the object.

This means that there are three main types of word order among the world's languages: SVO (subject-verb-object) sentences such as occur in English ("John eats a sandwich"); VSO sentences such as occur in Welsh ("Edrychais i ar lawer o bethau," "I looked at lots of things," literally: "Looked I at lots of things"); SOV sentences such as occur in Japanese ("Watashi wa hako o akemasu," "I open the box," literally: "I box open").

In languages where the descriptive adjective follows the noun, there may be a small number of adjectives that usually precede it, but in languages where the descriptive adjective goes before the noun there are no exceptions.

In French, most adjectives go after the noun "le livre rouge" ("the red book," literally: "the book red"), but there are indeed a small number of adjectives that usually precede the noun as in the sentence "Je vois le petit enfant" ("I see the small child"). In contrast, English sentences have the adjective before the noun and never after (except perhaps in poetry).

If a language has gender categories for nouns, it has gender categories for pronouns.

Many languages use gender, such as French "le" (the; masculine) and "la" (the; feminine) or Norwegian "en" (a; masculine), "en/ei" (a; feminine), and "et" (a; neuter). Languages without noun gender include Finnish, which does not have gender categories for its pronouns either, using the word "hän" for both "he" and "she."

In the time since Greenberg published his survey, other linguists have found hundreds of additional universal characteristics among the world's languages such as:

There exist eleven basic color terms: black, white, red, green, blue, yellow, brown, purple, pink, orange, and gray. Every language describes colors as a mixture, shade, or subcategory of one or more of these eleven basic color terms.

If a language has a distinct term for "foot," it will also have a distinct term for "hand"; if there are terms for "fingers," there will also be terms for "toes."

Languages tend to incorporate the following semantic distinctions into certain words: dry/wet (e.g., "dirt"/"mud"), young/old (e.g., "foal"/"nag"), alive/dead, long/short, male/female, light/dark, et cetera.

Another way to look at language universality is to see whether speakers of different languages share similar or identical associations for certain words, both concrete (such as "table") and abstract (such as "love"). In 1961, the American psychologist Mark R. Rosenzweig performed a survey of word associations with a group of American, French, and German students, which demonstrated remarkable similarities of associations across the languages. For example, given the stimulus word "man" (or its equivalent in French and German), the most frequent response for all three groups was "woman"; for "soft" the most frequent response across all three groups was "hard"; for the abstract "sickness" the most common response for each group was "health." Even when the most common response for each group of students was not the same, it was very similar: given the stimulus word "bath," the most common German response was "water"; for the French group it was "sea" and for the American group, "clean."

Finally, universality throughout the world's languages can be observed not only in grammar and words, but also in the subjects about which speakers talk with one another. The anthropologist Donald E. Brown has sifted through ethnographical archives for all documented cultures from around the globe, searching for universal patterns of behavior. His research (entitled "Human Universals") indicates a stunningly high level of similarity across thousands of cultures, from the Inuit and

Samoan to the English and American. Among the hundreds of universals Brown found are:

Antonyms: word pairs that are opposite in meaning such as "hot"and "cold."

Baby talk (also known as "motherese" or "mamanaise" in French): a form of speech used by adults when talking to babies or toddlers, usually high in pitch and delivered with a "cooing" pattern of intonation.

Classifications of age, body parts, colors, fauna, flora, emotions, kin, sex, space, tools, and weather.

Folklore.

Gossip.

Humorous insults and jokes.

Linguistic redundancy: the repetition of information in a message so that it becomes increasingly unlikely for errors to occur in its transmission or reception. English, for example, has been calculated to be 75 percent redundant, meaning that English sentences are 75 percent longer than if the alphabet were used as efficiently as possible for encoding messages.

Logical notions of "and," "equivalent," "general/particular," "not," "opposite," "part/whole," "same."

Making comparisons.

Metaphor.

Metonyms: the use of a single characteristic to identify a more complex entity (such as "sweat" for hard work and "tongue" for language).

Numerals: at least "one," "two," and "more than two."

Onomatopoeia: a word that imitates the sound it is describing (such as the English words "buzz" and "click").

The past/present/future.

Personal names.

Poetry/rhetoric.

Possessive: as in the English "John's car" or "That book
   is mine."
Pronouns: I, he, we, they, et cetera.
Proverbs, sayings.
Special speech for special occasions.
Symbolic speech/conduct.
Synonyms: words with identical or similar meaning.
Time: cyclicity of, units of.

Brown has used his research to urge his fellow anthropologists to seek out the things that unite, as well as separate, human cultures from one another: "Humans are and must be sensitive to differences. But too much focus on difference lurks behind human conflict. We should find hope in realizing how rich and numerous our commonalities are."

Learning a Second Language

Considering what we know about the universality or near-universality of many characteristics (grammar, word associations, concepts) found across the world's languages, it is puzzling that so many foreign language learners should find acquiring their target language an often difficult and frustrating experience. Why might this be?

 Many people have read or heard that there is a "critical period" during childhood for learning language, and that learning a second language as an adult constitutes a considerable handicap. However, this assertion is based on a misunderstanding of what language scientists mean when they talk about "language windows" or "critical periods." In fact, these terms refer to the acquisition of a first language as a young child, and not to the learning of subsequent languages later in life.

 The concept of a "critical period" for acquiring language in early childhood was popularized by linguist Eric Lenneberg in the 1960s. According to Lenneberg, the first years of childhood are critical to the successful acquisition of a first language. If language input does not occur until after this time, the individual will never achieve a full command of language, particularly grammar (sentence formation).

An example of this comes from the case of "Genie," who spent the first thirteen years of her life locked inside her room by her mentally unbalanced father. Discovered by authorities and taken into foster care as a teenager, she was almost completely mute, with an extremely limited vocabulary of around twenty words. Genie spent the following years working intensively with language scientists, during which time her vocabulary grew considerably, but she remained unable to produce grammatically correct sentences. For example, asked to create a question, she responded: "What red blue is in?"

The extent to which the "critical period" affects second language learning remains a matter of considerable debate among linguists. Though it seems true enough that few adult language learners acquire the native-like fluency that young learners display, a small percentage of adult bilinguals do achieve comparable levels of fluency even though they did not begin learning the language until adulthood. For this reason, many linguists accept that learning a second language is not subject to a biological "critical period."

A recent finding by researchers at the University College London Centre for Human Communication demonstrates that, given the right stimulus, the adult brain can indeed be retrained to accurately acquire the sounds of a second language. It is well established that adult learners can struggle to distinguish certain sounds in a new language: German-speaking students trying to learn English, for example, often find it difficult to distinguish between the "v" in "vest" and the "w" in "west." Studies by linguists Paul Iverson and Valerie Hazan examined whether it is possible to retune how the adult brain processes sounds in a second language. In one study, Japanese subjects were retrained to hear the difference between the sounds "r" and "l" (something that Japanese students of English find especially difficult). Before and after training, the subjects received a series of perceptual tests to evaluate their perception of acoustic cues. By the end of the ten-week training period the subjects had improved their recognition of the two sounds by an average of 18 percent.

Dr. Iverson's conclusion from these studies is optimistic for adult language learners:

> Adult learning does not appear to become difficult because of a change in neural plasticity . . . learning becomes hard because experience with our first language "warps" perception. We see things through the lens of our native language and that "warps" the way we see foreign languages . . . we change our perception during childhood so that it becomes specialized to hear the speech sounds in our first language. This specialization can conflict with our ability to learn to distinguish sounds in other languages. Through training, we can essentially change our "perceptual warping" to make second-language learning easier.

Other research suggests a possible reason for the extra difficulties that face learners acquiring a second language as adults, compared to those who learn it in childhood. A 1997 study by neuroscientists Joy Hirsch and Karl Kim found that second languages are stored differently in the human brain depending on when they were learned.

Dr. Hirsch recruited twelve bilingual subjects (ten different languages were represented in the group). Half had learned two languages as infants; the other half had begun learning their second language around the age of eleven and had acquired fluency by nineteen after living in the country where the second language was spoken. The subjects were asked to produce complex sentences describing what they had done the previous day, first in one language and then the other. As they spoke, a functional

MRI tracked blood flow within each subject's brain to help find out where in the brain thinking in each language occurred.

Several areas of the brain are specialized for language, such as Wernicke's area, a region devoted to understanding the meaning of words and oral comprehension. Another is Broca's area, which is dedicated to speech and some deep grammatical elements of language. In the bilinguals who had learned both languages in early childhood, activity in their brain's language areas was identical for both languages. But when the people who had learned their second language in adolescence were scanned, their brains showed the Broca's area divided into two distinct areas, with one area activated for each language. Though the two areas were close together, they were always separate.

This result suggests that, when the learning is early, the brain treats multiple languages as a single language, whereas a second language acquired later in life is treated as distinct by the brain and for this reason stored separately. This implies that the brain uses different strategies for learning languages, depending on age. For example, a baby learns to talk in an especially stimuli-rich environment, using a range of senses—hearing, vision, touch, and movement—that feed into hardwired circuits of the brain including Broca's area. Once the cells become tuned in to one or more languages, they become fixed. Two languages learned at this time become intermingled. People learning a second language later in life have to acquire new skills for generating the complex sounds of the new language, because Broca's area is already dedicated to the native tongue; so, an additional area of the brain is recruited.

Neurosurgeons have also found evidence that adds weight to the idea that multiple languages are stored separately in the brains of adult learners. Dr. George Ojemann, a professor of neurology at the University of Washington School of Medicine, operates on people who suffer severe epileptic seizures. Dr. Ojemann performed a study with bilingual patients in which he showed each a picture of an everyday item (such as a banana) and asked him to name it. By using very precise electrical stimulation of specific brain regions, Ojemann could get the patient to talk, for example, in Spanish but not English, then stimulate a nearby area and get the opposite result.

Another striking example of how the brain stores languages learned at different times of life in different areas of the brain comes from a conversation I had with a neuroscientist while doing research for this chapter. He recounted the case of a young Vietnamese woman who had been brought up with two languages—Vietnamese and English—and had for the past few years as an adult begun to learn some French. Following a car accident, she lost her ability to use either of her native tongues, but her French was preserved. Each time that her husband, who was also originally from Vietnam, came to visit her at the hospital, the neuroscientist—a French-English bilingual —was required to serve as an interpreter so that the couple could converse together.

Though scientists still have a way to go in understanding how the brain stores and processes language, there is growing consensus that learning a second language, whether in childhood or as an adult, is undoubtedly good for you. Suzanne Flynn, a professor of linguistics and second-language acquisition at the Massachusetts Institute of Technology, states that people who speak two languages have a "distinct advantage" over monolinguals because from an early age bilinguals are "better able to abstract information . . . they learn early that names of objects are arbitrary, so they deal with a level of abstraction very early."

Psychologists use a simple game consisting of Lego and Duplo blocks to evaluate differences in perception between monolingual and bilingual four-year-olds. First, towers are built using either the Lego or Duplo blocks (the Duplo blocks are just like the familiar Lego ones except they are much bigger). Then the psychologist tells the children that each block, regardless of its size, holds one family. The children's task is to look at a tower and say how many families it can hold. Answering the question correctly depends on the child's ability to ignore the visual fact that a tower made of seven Lego blocks is the same height as a tower made of four Duplo ones. While monolingual children can do this by age five, bilingual children can do it at four. It appears that the bilingual children are better able to focus attention and ignore distractions.

Knowing more than one language may even help ward off age-related declines in mental performance. In studies performed in Canada, India, and Hong Kong, psychologists determined that bilingual individuals performed better than monolinguals on tests that measured performance speed on a task while distracted. The researchers used the Simon task, a test used to measure mental abilities known to decline with age. Subjects view a flashing red or blue square on a computer screen and press either a left- or right-side button, depending on which color appears. Three experiments showed that bilingual speakers of Cantonese and English, Tamil and English or French and English consistently outperformed subjects who only spoke English. The researchers hypothesize that the ability to hold two languages in the mind simultaneously, without letting words or grammar slip from one into the other, might account for the greater control needed to perform well at the Simon task. Another possibility is that bilinguals have better working memories for storing and processing information.

Successful Language Learning

Learning a new language involves three major steps: acquiring the sounds (phonology) of the new language; learning new words, their meanings, and the rules governing word formation (morphology); and knowing how to use the words in grammatically correct combinations to form phrases and sentences (syntax). A useful definition for language acquisition comes from the excellent *An Introduction to Language* by linguists Victoria Fromkin and Robert Rodman: "Knowing a language means being able to produce new sentences never spoken before and to understand sentences never heard before."

One reason adult language learners find mastering the sounds of a second language so much harder than their first is "transfer," the interference by previously learned knowledge of new information we acquire. This is why many adult learners tend to pronounce the sounds of their first language while attempting to speak a second. For example, English speakers tend to produce a puff of air (the linguistic term is "aspiration") when saying words such as "come" and "quick," which English learners of Spanish often produce in similar-sounding words such as "como" and "que" even though Spanish speakers do not aspirate in this way. Of course, the same problem in reverse affects many Spanish speakers learning English.

A method I use for practicing the sounds of a target language is to construct sentences that repeat each sound several times over in quick succession, such as in the French sentence: "Alain tient du pain aux grains dans sa main" ("Alan holds granary bread in his hand").

Another helpful idea is to listen to songs in the target language and practice singing along to the lyrics. Using music in this way helps the learner imitate the language's rhythm and contributes to good pronunciation.

Part of pronouncing words accurately is knowing where to place the stress in each word. English is a very complicated language when it comes to stress: the "White House" has the stress on the word "white" whereas "a white house" has the stress on the word "house." Fortunately, many languages have regular stress: Finnish words always have the stress placed on the first syllable, whereas in Polish it is on the next-to-last syllable.

The way that a sentence is spoken (its "pitch contour") can also affect its meaning. Intonation languages, such as English and French, use pitch to change a sentence from a statement into a question. The sentence "John is here" (in French "Jean est ici") spoken with a falling pitch is a statement, whereas said with a rising pitch it becomes a question: "John is here?" ("Jean est ici?").

A language's phonology is not defined solely by an inventory of speech sounds, but by how a language assembles those sounds into syllables and the syllables into words. For example, the word "spletch" is not a word in English but could be, whereas "tskhviri" (Georgian for "nose") could not. Knowing which words are possible or impossible is second nature to a language's native speakers but must otherwise be learned consciously. One method for native English speakers is to study how English words that have been borrowed by the target language are written (and spoken) in it. For example, in Japanese (which does not allow consonant clusters) the English words "ice cream" and "girlfriend" have been imported as "aisukurimo" and "garufurendo."

Knowing how sounds combine in a language is also important for comprehension. For example, an English speaker hearing "thisroad" knows it must be "this road" and not "the sroad" because the combination "sr" is not possible in English. According to what linguists call "cohort theory," for each word spoken in a conversation the listener immediately and unconsciously narrows down a mental list of possible words (the "cohort") to determine which one he is hearing. For example, when he hears a word beginning with the sound "el-" his mind will automatically think of all known words beginning with that sound (elephant, electric, election, element, Eleanor), rapidly reducing the possibilities as the word's sounds unfold: "el-ec'" (electric, election) until the word is recognized: "el-ec-tion." Several experiments have supported this predictive form of word recognition, such as those that show that the recognition of a word is much more impaired by the mispronunciation of the initial letter of a word than by the mispronunciation of the final letter (because it is much easier to predict a word based on its initial portion than by its final one).

Learning vocabulary is often a daunting task for many second-language learners, but acquiring a rich, expressive vocabulary in a second language is not as hard as you might think. In the pages ahead, I will share with you a number of simple methods that will help you to learn new words quickly and effectively. Before I do, though, I want to tackle a common myth concerning vocabulary learning: that it is possible to "get by" in a language by learning a highly restricted vocabulary comprised of the language's one hundred most common words. For example, in his book, *Use Your Memory*, author Tony Buzan claims that just one hundred frequently used words make up 50 percent of everyday language. Here is an example of such a list for English:

a, an	after	again	all
almost	also	always	am
and	because	before	big
but	can	come	either
father	find	first	friend
from	go	good	good-bye
happy	have	he	hello
here	how	I	if
ill	in	know	last
like	little	love	make
many	mother	more	most
my	new	newspaper	no
now	of	often	on
only	or	other	our
out	over	person	place
please	same	see	small
some	sometimes	still	such

tell	thank you	that	the
their	them	then	there
there is	they	thing	think
this	time	to	under
up	us	use	very
want	we	what	when
where	which	who	why
with	yes	you	your

The problem with applying this idea to language learning is that the words listed are very basic indeed and carry very little information by themselves. As an example, here is the opening from a well-known children's book using only the one hundred basic words:

___ ___ ___ to ___ very ___ of ___ ___ ___ ___ on the ___, and of ___ ___ to ___: ___ or ___ ___ ___ ___ ___ the ___ ___ ___ ___ ___, but ___ ___ no ___ or ___ in ___, "And what ___ the use of a ___," " ___ ___ ` ___ ___ or ___?" ___ ___ ___ ___ in ___ ___ ___ (___ ___ ___ ___ ___, ___ the ___ ___ ___ ___ ___ very ___ and ___), ___ the ___ of ___ a ___-___ ___ ___ ___ the ___ of ___ up and ___ the ___, when ___ a ___ ___ with ___ ___ ___ ___ ___ ___. There ___ ___ ___ very ___ in that; ___ ___ ___ think ___ ___ very ___ out of the ___ to ___ the ___ ___ to ___, " ___ ___! ___ ___! I ___ ___ ___!" (when ___ ___ ___ over ___, ___ ___ to ___ that ___ ___ to have ___ ___ this, but ___ the time ___ all ___ ___ ___); but when the ___ ___ ___ a ___ out of ___ ___ ___, and ___ ___ ___, and then ___ on, ___ ___ to ___ ___, ___ ___ ___ ___ ___ ___ that ___ ___ ___ ___ ___ a ___ with ___ a ___ ___, or a ___ to ___ out of ___, and ___ with ___, ___ ___ ___ the ___ after ___, and ___ ___ ___ in time to see ___ ___ ___ a ___ ___-___ under the ___.

And now the full text:

Alice was beginning to get very tired of sitting by her sister on the bank, and of having nothing to do: once or twice she had peeped into the book her sister was reading, but it had no pictures or conversations in it, "And what is the use of a book," thought Alice, "without pictures or conversation?"

So she was considering in her own mind (as well as she could, for the hot day made her feel very sleepy and stupid), whether the pleasure of making a daisy-chain would be worth the trouble of getting up and picking the daisies, when suddenly a White Rabbit with pink eyes ran close by her.

There was nothing so very remarkable in that; nor did Alice think it so very much out of the way to hear the Rabbit say to itself, "Oh dear! Oh dear! I shall be late!" (when she thought it over afterwards, it occurred to her that she ought to have wondered at this, but at the time it all seemed quite natural); but when the Rabbit actually took a watch out of its waistcoat pocket, and looked at it, and then hurried on, Alice started to her feet, for it flashed across her mind that she had never before seen a rabbit with either a waistcoat-pocket, or a watch to take out of it, and burning with curiosity, she ran across the field after it, and fortunately was just in time to see it pop down a large rabbit-hole under the hedge.

Though it might be pointed out that some of the words in the story, like "waistcoat" and "daisy," are relatively rarely used, many others that do not make the "hundred words" list appear frequently such as "was," "get," "by," "her," and "it." Consider as well the absence from the list of huge quantities of other extremely common words such as "man" and "woman," "cat" and "dog," "hand" and "foot." What is more, the statistical argument—that one hundred words constitute 50 percent of everyday language—is misleading. To show this, an analysis of Shakespeare's *Macbeth* shows that 21 percent of the text is composed of just nine words (the, and, to, of, I, a, Macbeth, that, in) but this does not mean that an English learner knowing only these few words could understand one-fifth of the play's themes, events, or story line.

You simply cannot acquire a genuinely expressive and useful vocabulary in a target language by statistical lists; you need a range of ideas and methods. Learning onomatopoeic words is a good start because they occur in all languages and are naturally memorable: French "boum" (bang), German "klatschen" (to clap), and Spanish "susurro" (whisper) are examples.

A subtler form of onomatopoeia known as phonesthesia (where certain sounds become associated with particular meanings) is another way to learn and remember new words. An example of phonesthesia in English is the "gl" sound and its association with light: glow, glimmer, glitter, glint, gleam, et cetera. In Icelandic, words beginning with "hn" often describe round objects such as "hnöttur" (orb or globe), "hné" (knee), "hnúi" (knuckle), "hnappur" (button), "hnútur" (knot), and "hnipra" (to curl up into a ball).

A number of experiments by linguists offer further evidence for the idea that some words appear to be a more natural "fit" than others for the things they describe. In a 1954 experiment, the German psycholinguist Heinz Wissemann asked a group of subjects to invent words for various sounds. He found that the subjects tended to create words beginning with "p," "t," or "k" for abrupt sounds and words beginning with "s" or "z" for flowing sounds. In a more recent experiment involving natural language, the linguist Brent Berlin provided English speakers with fish and bird names from the Huambisa language (spoken in Peru). He found that they were able to distinguish the words for fish from those for birds significantly more often than chance, even though Huambisa bears no resemblance to English.

Test your own intuitive sense for word meanings from a range of languages with the following multiple-choice questions:

1. Does the adjective "pambalaa" in the Siwu language of Africa describe (a) a round, fat person or (b) an angular, thin person?
2. Is the word "durrunda" Basque for (a) a quiet or (b) loud noise?
3. Do the Japanese colors "aka" and "midori" mean (a) red and green or (b) green and red?
4. The Malay verb "menggerutu" refers to someone who (a) laughs or (b) grumbles?
5. Is the Italian "piro piro" a kind of (a) fish or (b) bird?
6. Do the Hungarian adjectives "nagy" and "kicsi" mean (a) big and small or (b) small and big?
7. If a Samoan says "ongololo" is he talking about a (a) centipede or (b) an ant?
8. In the Aboriginal Yir-Yoront language of Australia does the word "chichichi" refer to a dog that is (a) sitting or (b) running?

Answers: 1a, 2b, 3a, 4b, 5b, 6a, 7a, 8b

Another way of making sense of a language's vocabulary is to find relationships between its words. My naturally associative style of thinking lends itself especially to this form of learning. For example, when I think of the French word "jour" (day), I immediately think of "bonjour" (hello, literally, "good day") and "journal" (newspaper, daily); for the German word "Hand" (hand), I think of "Handy" (mobile phone) and "Handel" (a craft/trade). You can also make use of word clusters within a language to make the vocabulary more meaningful and memorable. An English example would be "pen," "paper," "pencil," and "paint," where all the words begin with the letters "pa" or "pe" and refer to similar objects, or those normally used together. In French, the words for "shoe," "sock," and "slipper"—all words for items worn on the feet—are strikingly similar: "chaussure," "chaussette," and "chausson," respectively.

Seeing the relationships between words in your native language and the target language can also be very helpful in remembering new vocabulary, though such connections are not always immediately apparent. For instance, the French word "ciel" (sky) can be related to the English "celestial." Below are some more examples from a range of languages:

Finnish: "joulu" (Christmas), Yule; "apteekki" (pharmacy), apothecary

French: "mer" (sea), marine; "doigt" (finger), digit; "coeur" (heart), coronary

Gaelic: "uisge" (water), whiskey; "sgrìobh" (write), scribble

German: "Vogel" (bird), fowl; "Herbst" (autumn), harvest

Lithuanian: "dantys" (tooth), dental; "senis" (old man), senile; "vyras" (man), virile

Romanian: "birou" (desk, office), "bureau; "ap " (water), aqua

Spanish: "año" (year), annual; "azul" (blue), azure

Welsh: "pen" (head) and "gwin" (white), penguin; "march" (horse), mare; "talu" (pay), toll

Of course, not all similar-looking words share the same meaning—such misleading resemblances are called "false friends" and include the German word "Gift"(which actually means "poison"), the Spanish "contestar" (meaning "to answer"), the Icelandic word "ský" ("cloud"), and the Finnish "kaniini," which, though resembling the English word "canine," actually means "rabbit."

A further way to use associations to remember vocabulary is to learn words that are used together frequently, "binomial expressions," such as the English "bread and butter," "bucket and spade," and "table and chair." Examples from French include: "mari et femme" (husband and wife), "poivre et sel" (salt and pepper), and "chien et chat"(cat and dog). Bear in mind that these expressions are sometimes expressed in reverse order to the English, such as the preceding two French examples (literally: "pepper and salt" and "dog and cat").

Compound words are a good source of vocabulary for the language learner, rather like "learn one word, get two free." When, in 2004, I was challenged by producers to learn conversational Icelandic in just one week before a TV interview in Reykjavik conducted entirely in the home language, I found that compounds were common and this helped me to extend my vocabulary in the language quickly. Examples include: "járnbraut" (railway) consisting of "járn" (iron) and "braut"(way); "hvítlaukur" (garlic), made up of "hvít (ur)" (white) and "laukur" (onion), and "orðabók" (dictionary), which is a combination of the words "orð" (word) and "bók" (book). Some compounds can even form part of other compounds, such as the Icelandic "alfræðiorðabók" (encyclopaedia) which adds the words "al" (all) and "fræði" (knowledge) to the compound word for dictionary. Using the expressive power of compounds, I was even able to create my own words, after just a few days' study of the language, that were readily understood by native Icelanders, such as: "orðafoss" (the experience of feeling immersed in a foreign language —literally "word waterfall") and "bróðurmál" (a second language that you feel a particular affinity towards— literally "brother tongue").

Learning a language's affixes—sounds that attach to the beginning or end of a word to create new meanings—will also enrich a learner's vocabulary. An example from French is the suffix "-ier" which when used with the words for various fruits or nuts refers to the tree or bush that produces it: "pomme" (apple), "pommier" (apple tree); "groseille" (red currant), "groseillier" (red currant bush); "châtaigne" (chestnut) and "châtaignier" (chestnut tree). Another example, this time from Spanish, shows how affixing a single noun can sometimes create a wealth of new meanings: "manteca" (lard), "mantequilla" (manteca + illa, "butter"); "mantequero" (manteca + ero, "butter dish"); "mantecada" (manteca + ada, "buttered toast and sugar"); and "mantecado" (manteca + ado, "butter cake")..

It is also possible to increase your expressive capacity in a second language by learning the range of possible meanings for certain common words, depending on their context. In French, for example, the word "seul" can be used for both the meanings "alone" and "only," while the word "toujours" can mean either "always" or "still" (as in "Il est toujours au chômage": "He is still unemployed"). Another example is the German word "Schuld," which can be used to mean either debt, fault, responsibility, or guilt. Beware, however, of directly translating a word with multiple meanings from your native language into a second language, as this rarely works. For example, the English verb "work" can be used in various senses such as "She works as a teacher" and "The elevator does not work," but in French the verb is different for each meaning : "Elle travaille comme professeur" but "L'ascenseur ne fonctionne pas. "

A particular bugbear for many language learners (especially those whose native language is English) is the use of grammatical gender (the assignment of gender to all nouns) in many languages. Most European languages have two or three genders (such as the German "der" for masculine nouns, "die" for feminine nouns, and "das" for neuter nouns), though that number pales in comparison to the aboriginal Yanyuwa language, which has no fewer than sixteen noun classes based on the various functions of objects used in their society! One reason why learning a noun's gender is often so tricky is its seeming arbitrariness; for example, in French the word for "moon" (la lune) is feminine but in German it is masculine ("der Mond"). Mark Twain, the American humorist, marveled at the gendered nature of German nouns in his book, *A Tramp Abroad*: "In German a young lady has no sex, while a turnip has . . . (A) tree is male, its buds are female, its leaves are neuter; horses are sexless, dogs are male, cats are female . . . tomcats included."

Studies by cognitive psychologists Lera Boroditsky, Lauren A. Schmidt, and Webb Phillips suggest that native speakers of languages with gendered nouns remember the different categorization for each by attending to differing characteristics, depending on whether the noun is "male" or "female." In one such study, a group of native German and Spanish speakers was asked to think of adjectives to describe a key. German speakers, for whom the word "key" is masculine, gave adjectives such as "hard," "heavy," "jagged," and "metal," whereas the Spanish speakers, for whom "key" is feminine, gave responses like "golden," "little," "lovely," and "shiny."

Another way for learners to remember the gender of a noun is to learn each noun with its article ("a" or "the") and/or an adjective that reflects the noun's gender. For example, it is far easier to remember that the French noun "coccinelle" (ladybird) is feminine if you learn it as part of a phrase like "la petite coccinelle" (the little ladybird). In many languages, a noun's gender can also be guessed from its ending. In German, nouns ending in "-chen" or "-lein" are always neuter ("das Mädchen," girl) while those ending in "-ik" are often feminine ("die Klinik," "die Musik," "die Panik"), and those ending in "-ismus" are masculine ("der Journalismus," "der Kommunismus").

The final stage in language acquisition is putting the words together in meaningful chunks (phrases) that can be combined into longer sequences of words (sentences). Reading in the target language is an essential component of this process for me, as learning the words within a context aids memory and helps me to use them correctly in my own sentences. For my weeklong study of Icelandic, I read as much as possible, quickly learning many useful word chunks with which I could build my own sentences. For example, below are five short sentences in Icelandic:

"Ég las góða bók um helgina."
("I read a good book at the weekend.")

"Hún er alltaf með hatt á höfðinu."
("She is always with a hat on her head.")

"Ég heyri fuglana syngja."
("I hear the birds singing.")

"Stundum kemur regnbogi þegar þad rignir."
("Sometimes a rainbow appears when it is raining.")

"Hann fer á fætur klukkan sjö."
("He gets up at seven o'clock.")

By recombining word chunks from these sentences, I can immediately create the following original sentences (among others):

"Stundum heyri ég fuglana syngja klukkan sjö."
("Sometimes I hear the birds singing at seven o'clock.")

"Hún er með góða bók."
("She has a good book.")

"Það rignir alltaf þegar ég fer á fætur."
("It is always raining when I get up.")

"Hann kemur um helgina."
("He is coming at the weekend.")

The more sentences you read in the target language, the more combinations of words and the more grammatical patterns you will be exposed to, with the advantage that you can start creating your own complete, grammatically correct sentences almost immediately, without recourse to verb tables or lists of grammatical rules.

Learning a language in this way avoids the problems that often arise from trying to learn the words one by one, outside of any context. An example of such a problem is when otherwise very capable English learners make mistakes like "the ripe man" (instead of "the mature man") and "the sea was profound" (instead of "deep"). In many languages, even everyday expressions are formed with important grammatical differences from their English equivalents, which are therefore best learned through examples, like the following French ones:

"Le garçon s'est lavé les dents."
(Literally: "The boy to him washed the teeth.")

"Tu me manques."
(Literally: "You to me lack," "I miss you.")

"Il fait froid."
(Literally: "It makes cold," "It is [the weather] cold.")

"J'ai mal à la tête."
(Literally: "I have pain at the head.")

Learning whole phrases and sentences, and the vocabulary in various combinations within them, works so well because it mimics the process by which we all learned our first language. After all, no child acquires his mother tongue by studying its grammar or making lists of words. A word-by-word approach to learning a second language is unnatural precisely because it is too fragmentary: we learn to speak a language, not to reproduce the vocabulary we have read in a book or heard on a cassette, but to communicate our own thoughts, feelings, experiences, and ideas—all things that require complex, well-turned sentences.

Using ideas and methods like those I have outlined in this chapter, I believe it is possible for anyone to successfully acquire a second language and, just as important, to draw pleasure and insight from the experience along the way.

Vanishing Voices

I know well the practical and personal value that comes from accessing the wide array of knowledge and understanding available from our linguistically rich and diverse world. For this reason, I am deeply shocked and saddened by reports that as many as a half of the world's approximately six thousand languages are seriously endangered. According to some estimates, as many as 90 percent of the world's languages are likely to disappear during this century. The linguist Michael Krauss calls "catastrophic" this potential loss of linguistic diversity, arguing that it "should be as scary as losing 90 percent of the world's biological species."

Here is just one example: as of this writing an elderly woman named Marie Smith Jones has died, the last speaker of the Alaskan Native Eyak language. Working with endangered language researchers, Ms. Jones became a champion of indigenous rights and conservation, helping the University of Alaska to compile an Eyak dictionary in the hope that future generations would resurrect it. Ms. Jones's daughter reports that none of her mother's nine children grew up to speak Eyak because they grew up "at a time when it was considered wrong to speak anything but English."

The historical perception of minority languages as inferior or uncivilized is one reason behind the striking decline in the numbers of speakers for many of the world's languages. In the United States and Canada, Native American languages were historically regarded as "barbarous" and school students were punished harshly if they spoke in their native language. Languages, like plants and animal species, also disappear when the natural habitats of their speakers are destroyed, whether by population pressures, genocide, or the spread of industrialization. Other reasons include political oppression—such as that experienced by my Lithuanian friends during the Soviet era, when newspapers and road signs were produced only in Russian—and the spread of foreign-language electronic media, described by one linguist as "cultural nerve gas."

Of course, some language loss, like species loss, is natural and predictable. Just like plants and animals, languages grow, spread, and change over time before eventually dwindling and dying. The moral and practical issues surrounding language preservation are complex; as the inventor of the international language Esperanto Ludwig Zamenhof noted, linguistic differences between communities can be divisive and a source of misunderstanding, tension, and even outright hostility. It is not hard to see why many young people from minority language communities prefer to speak a major European language, or Russian, Arabic, or Chinese, that promises economic and social advancement.

There are good reasons, however, why we should all care

about the plight of endangered languages. Languages are not only groupings of words, but also the means for the transmission of a culture that has adapted to unique environmental, sociological, and historical circumstances. Words describing a particular idea or cultural practice rarely translate well into another language. Many endangered languages possess rich oral cultures of songs and stories that risk being lost forever. Unique natural and scientific knowledge is bound up in many of these languages. Indigenous groups that have interacted closely with the natural world for hundreds or thousands of years encode knowledge about local plants, animals, and ecosystems within their languages—much of which has yet to be documented by scientists. The study of these languages by researchers therefore aids environmental and conservation efforts, and increases our understanding of how humans store and communicate knowledge.

Being an optimist, I believe there is still hope for our many diverse languages. For one thing, there are examples of extinct or dying languages coming back to life again. In Britain the once-dead languages of Manx and Cornish have been successfully revived, with some preschool and primary school children on the Isle of Man taught entirely through the Manx language. After decades of dwindling numbers of speakers, Welsh recently showed its first increase in speakers in generations. The most successful language revival by far has been that of Hebrew, which after two millennia was revived during the nineteenth century and is today the official language of the state of Israel, spoken by more than seven million people.

The late Massachusetts Institute of Technology professor of linguistics Ken Hale observed that, "The loss of language is part of the more general loss being suffered by the world, the loss of diversity in all things." Being just as passionate for languages as Hale, I agree wholeheartedly. In a rapidly shrinking world, we must take care not to squeeze out many of the unique tongues through which people have for generations perceived, imagined, thought, and talked about the nature and meaning of the world and their place within it. Wisdom is only possible when we choose to listen and to learn, and for this reason every voice—in every language—should have the right to express itself and be understood.

5

The Number Instinct

A s a child, I was fascinated by the beauty, order, and complexity of numbers. Two decades on I remain as entranced as ever, but I have become equally as interested in the processes that give rise to my idiosyncratic mental experience of them. In my head, numbers assume complex shapes that interact to form solutions to sums, but how exactly do I visualize and compute numbers? And does my unusual ability derive, at least in part, from a "number sense" that everyone shares?

In the following pages I marshal considerable evidence to support the idea that most people are born with a hard wired capacity to count, analogous to the "language instinct" examined in the previous chapter. This inborn "number instinct" manifests itself in many forms—from our shared systems of arithmetic to personal internal representations of the world of numbers. My own theory of savant numerical abilities is that they are the result of purely natural (i.e., not computerlike) processes within the brain. In fact, if I am right, my mathematical abilities are the by-product of a form of mental computation that most people do every day, both spontaneously and effortlessly.

Everybody Counts

Only two years out of university, infant-cognition researcher Karen Wynn stunned the academic world when she published the findings of her remarkable study showing that five-month-old babies possess intuitive counting skills. So significant did her results prove that they subsequently inspired a wave of studies into humans' seemingly innate "number sense."

Wynn's ingenious method for evaluating young babies' arithmetic abilities involved screens, Mickey Mouse dolls, and her subjects' curiosity. In one task, the five-month-olds watched an experimenter put a doll on a table and then place a screen in front of it. Then they watched the experimenter place a second doll behind the screen. The babies did not know that a second experimenter was hiding behind the screen, able to add or remove dolls. In half the experiments the screen was removed to show the correct number of dolls, but in the other half an incorrect number of dolls appeared (equivalent to $1 + 1 = 1$ or $2 - 1 = 2$, for example). The babies looked considerably longer at the incorrect number of dolls. Wynn concluded that this was because the babies, who naturally look longer at new or surprising items than at familiar ones, were able to count how many dolls were behind the screen and were baffled by the errant results.

Babies also responded to the numbers of actions and sounds, Wynn found. In one experiment, infants were shown a puppet that jumped two hops repeatedly. Later they were shown the same display, except that the puppet switched between two hops and three. The babies looked for longer at the novel number of jumps, indicating that the babies had counted how many jumps occurred in each sequence. In a similar experiment, seven-month-olds listened to dolls that emitted a sequence of two or four sounds. The babies looked significantly longer at the doll when it produced the sequence of sounds that they had not heard before.

Infants have also been shown to excel at tasks involving numbers considerably larger than 1, 2, and 3. Psychologists Elizabeth Spelke and Fei Xu performed a study with six-month-olds using arrays of dots with varying sizes and positions. Even though all the arrays covered exactly the same amount of space, the babies were able to distinguish between arrays of 8 and 16 dots. In a related study, kindergarten-age children were able to tell that an image of 21 dots followed by one with 30 dots contained more dots than appeared in a single image of 34 dots.

Spelke has recently extended these results further still, by showing that young children who know how to count can approximately add and subtract numbers even before they have learned the rules of arithmetic. The researchers gave groups of five- and six-year-olds a laptop-based audiovisual test that asked questions such as: Sarah has 15 candies and she gets 19 more; John has 51 candies. Who has more? Remarkably, the children answered correctly in 64 percent to 73 percent of trials.

If there is a separate, specialized "number instinct," similar to the Chomskyan language instinct, it should have its own location in the brain, distinct from general logical reasoning skills. Some medical cases, in fact, support this idea. Signora Gaddi, an Italian in her late fifties, suffered a stroke that damaged the left parietal lobe of her brain, causing her to become "acalculic," unable to use numbers. Subsequently studied by researchers, she was able to speak fluently and live independently (even running her own hotel business) but was unable to count past four, and only then by slowly counting each item one by one.

Further supporting the idea of a distinct "number instinct" are cases of people with good numerical abilities but very poor general cognitive functioning. Mr. Bell, a patient at London's National Hospital for Neurology and Neurosurgery, suffered from a degenerative brain disorder called Pick's disease. Mr. Bell's language had almost completely disappeared, except for a few stock phrases such as "I don't know" and, bizarrely, "millionaire bub." His understanding of speech or written text was also close to nonexistent. Yet he could still add and subtract accurately, perform multiplications, and tell which of a pair of two- and three-digit numbers was larger.

There are other hints at innate number instinct within us all. For example, as with language, there exist myriad forms of counting and arithmetic throughout the world's history and geography. Let's take a closer look at these to see what more they can tell us of the relationship between math and our minds.

Counting Around the World

In his book *The Mathematical Brain*, cognitive neuroscientist Brian Butterworth argues persuasively that an innate "number module" exists within the human brain. In support of his thesis, he points to the seemingly universal ability among humans to instantly recognize a quantity of between one and four items without needing to count them (a process known as subitizing, from the Latin "subitus" meaning "sudden").

Many languages provide evidence for an innate number module by treating the numbers one to four differently from larger ones, often giving them special grammatical forms that do not exist for the numbers five and above. One example of this is how some languages distinguish between a group of exactly two objects and any larger group of objects. The Scottish Gaelic language, for instance, uses the word "chloich" (stones) only when it is preceded by the number two (or larger numbers that end in a two, such as twenty-two); for all other numbers it uses the standard plural "clachan." English provides another example of such linguistic distinctions for the smallest numbers. We form the ordinal numbers (such as "seventh" or "eighty-ninth") in English by adding "-th" to the end of a number, yet the numbers one to three have special forms ("first," "second," and "third") that do not follow this rule.

It is the Icelandic language, though, which provides some

of the most impressive linguistic evidence for an innate bias towards the smallest numbers. In Icelandic, the numbers one to four are treated like adjectives, given a multitude of forms depending on how they are used in a sentence. Higher numbers, however, only have a single form—suggesting that they are not considered descriptive in the way that the lowest numbers are. For example, the word for "four" is always modified to agree with the object it describes (as are all adjectives in Icelandic), so "fjórir menn" (four men) but "fjögur börn" (four children) and "eitt af fjórum" (one in four). In contrast, the word "five" (and those for any other larger number) does not change: "fimm menn" (five men), "fimm börn" (five children), "eitt af fimm" (one in five). Such examples suggest that the brain perceives the lowest numerosities as tangible properties of groups of objects (e.g., perceives the "threeness" or "fourness" of something) in much the same way as it does an object's color, shape, or temperature.

Historical and anthropological evidence also point to a universal, innate capacity to count and use numbers. For example, the Sumerians and Babylonians used numerals from at least 3000 BC—clay tablets from the period have multiplication tables, square and cube roots inscribed on them. Astoundingly, tally marks have been found on pieces of bone and cave walls that date back thirty thousand years—showing perhaps that a counting instinct goes back as far as the human instincts for art and language.

Anthropologists have documented the use of counting words in virtually every culture around the globe. Though some of the most remote cultures have relatively few number words (typically for "one," "two," and sometimes "three," and "many"), they have devised various ways to count beyond these lowest values. For example, the Bushmen of South Africa count "xa" for one, "t'oa" for two, and "'quo" for three—to count four they use "t'oa-t'oa" (literally "two-two"); five is "t'oa-t'oa-ta" (two-two-one); and six is "t'oa-t'oa-t'oa" (two-two-two). Variants of this method have also been recorded in Southeast Asia, North America, Central Africa, and the Pacific.

To help them count higher than five or six, many people in Papua New Guinea count not only on their fingers, but also on their toes and other parts of the body. For example, the Oksapmin tribe use a counting system up to 27 starting with the right thumb and continuing up to the nose, which equals 14, eventually culminating at the left little finger. The words for each number are the names of parts of the body, so that "eight" in the Oksapmin language literally translates as "right elbow." The cultural anthropologist Geoffrey Saxe reported observing an Oksapmin trade store owner adding 14 and 7 shillings together by first counting up to the nose for fourteen, then continuing through the left eye (15), the right thumb (1), the left ear (16), the first finger (2), and so on until he arrived at the correct total (the left forearm).

Using the names of parts of the body to count is not restricted to remote cultures—some linguists argue that the English words "four" and "five" are related to the root of the word for finger. The Gothic word "fimf" (five) may also have the same origin ("figgrs" meaning finger). In Slavic languages the word for five is related to the word for fist. Research also suggests that the original Indo-European word for "ten" had the literal meaning "two hands."

A related side note: research suggests that the counting words we use in English (and many other European languages) can have a negative effect on some young children's numeracy and arithmetic skills. Studies consistently show that Asian children learn to count earlier and higher than their American counterparts and can do simple addition and subtraction sooner. The reason: the teen and ten numbers in English and other languages are irregular and difficult for children to learn. In contrast, the number words in most Asian languages are much more consistent; in Chinese the word for eleven is "ten one," twelve is "ten two," thirteen is "ten three," and so on. This pattern continues into the tens: twenty is "two ten," thirty "three ten," and forty-five is "four ten five." A sum like "eleven plus twelve equals twenty-three" is rendered as "ten one plus ten two equals two ten three" in Chinese. The language helps, rather than hinders, early understanding of the base 10 system.

That we perform arithmetic in tens (base 10) is likely the result of our possessing ten fingers and toes. Other bases, though. have been used by various cultures throughout history, indicating that counting and arithmetic were not the result of any single invention that diffused around the world, but rather of spontaneously created and diverse methods for making large numbers—further evidence for a universal number instinct.

English contains signs of some of these alternative bases, such as base 12 for counting money (12 pennies to the shilling in England), length (12 inches to a foot), and quantity (dozen). The English word "score"—meaning 20—is a trace of a vigesimal (base 20) system. In French the word for 80 is "quatre-vingts" (literally "four twenties") while in Danish the word for 60 is "tre-synds-tyve" (three times twenty). The ancient Mayan culture even had words for the first four powers of 20: "hun" (20), "bak" (20 x 20 = 400), "pik" (20 x 20 x 20 = 8,000) and "calab" (20 x 20 x 20 x 20 = 160,000). The Babylonians went further still, using a sexagesimal (base 60) system, which survives to this day in the way we measure angles and time. The base is very convenient for certain kinds of sums, because it has many factors: 2, 3, 4, 5, 6, 10, 12, 15, 20, and 30, which help simplify calculations.

The ancient Egyptians had their own simple, intuitive method for performing multiplications, requiring them only to add and double. For example, to solve the sum 42 multiplied by 14, we double 42 successively as follows:

$$
\begin{array}{ll}
1 & 42 \\
2 & 84 \\
4 & 168 \\
8 & 336
\end{array}
$$

We stop here because 8 doubled (16) would be larger than 14. As 8 + 4 + 2 = 14, we add the corresponding totals above together to arrive at the final answer: 336 + 168 + 84 = 588.

Division was also performed using this method. For the sum 143 divided by 11, the Egyptians would think of it as the question: "What multiplied by 11 equals 143?" and then double 11 successively as follows:

1	11
2	22
4	44
8	88

We stop here because the next doubling would produce 176, which is greater than 143. The numbers which make up 143 from the right side of the table are 11, 44, and 88 (11 + 44 + 88 = 143) and the numbers beside them added together give the answer to the sum: 1 + 4 + 8 = 13.

Though ancient, this method is still used today in rural communities as far afield as Ethiopia, Russia, and the Arab world.

Numbers in the Brain

Accepting the evidence for a number instinct within all of us, we might naturally ask: how are numbers represented within the brain? Do numbers, like certain words in language (such as those beginning with "sl-" in English (sloppy, slithering, slam), which seem to produce negative feelings), generate particular forms, feelings, or other associations in the human mind?

One of the most remarkable studies on how people mentally evaluate numbers took place in 1967 when researchers Robert Moyer and Thomas Landauer made a startling discovery. The men had asked a group of healthy adult subjects to decide which of two single-digit numbers (such as 7 and 5) flashed onto a screen was the larger, and carefully timed the subjects' response. They found that the adults often took more than half a second to reply to such simple comparison tasks, and even then made mistakes. Even more surprising, though, was the fact that when the two digits represented very different quantities, such as 2 and 9, the subjects responded quickly and accurately, but when the digits were much closer together, such as 5 and 6, the subjects were much slower and made mistakes in about 10 percent of trials. Even when the distance between the numbers was the same, response times slowed as the numbers got larger; that is, it took longer to evaluate which was larger of "7 or 9" than for "2 and 4".

A similar effect was observed when subjects were asked to compare two-digit numbers. The researchers found that the subjects took longer and made more errors when asked to name the larger number in a pair such as 71 and 65, compared to one such as 69 and 61 despite the fact that they would need only look at the first digit in each number of the first pair to recognize which was larger.

Why these differences in speed and accuracy for simple number-comparison tasks? Moyer and Landauer hypothesized that the subjects were converting the numbers onto a mental number line, which they used to make a comparison of the quantities. Other researchers have subsequently confirmed their findings.

There are several clues to how this mental number line is organized in the human mind. One is found when you ask people to think of a number at random between 1 and 50. Though it might be supposed that each number would be given in reply with equal probability around 2 percent (1 in 50) of the time, in fact, given a large enough sample of people, a systematic bias can be observed: people will tend to produce smaller numbers more frequently than larger ones. This suggests that numbers are mentally represented rather like the logarithmic scale on a slide rule where roughly equal space is given to the interval between 1 and 2, between 2 and 4, or between 4 and 8. Our "mental ruler," it would seem, compresses larger numbers into a smaller space, which is why smaller numbers are more accessible to our minds.

A further experiment shows that our minds represent numbers spatially. The subject sits at a computer, and is asked to respond to numbers that appear on the screen by pressing either a left or right key. Scientists have discovered that, on average, subjects respond more quickly to smaller numbers with the left key and to larger numbers with the right one. The same effect occurs even when the subjects cross their arms or use a single hand for both keys.

The neuroscientist Stanislas Dehaene, who first identified this spatial effect in 1993, theorizes that it is due to the fact that people spontaneously imagine the numbers on a mental number line where small numbers are always to the left. Dehaene also found that the effect depends on the subjects' culture; those who read from right to left (such as Iranians) exhibit the same effect but in reverse, with faster responses to smaller numbers with the right key and to larger numbers with the left.

Researchers have found that as many as 10–15 percent of people are able to visualize their mental number line, while a smaller number report that their numbers have colors, textures, or even personalities. Francis Galton, a psychologist and cousin of Charles Darwin, was the first to discover this phenomenon back in 1880. Galton circulated a questionnaire among friends and acquaintances, asking them to report if they could "see" numbers and if so in what way. The responses he obtained offer a fascinating glimpse into the sheer variety of people's mental number representations, though many of the number lines also shared certain characteristics: about two-thirds ran left-to-right and more often upwards than downwards. Some of the individuals' number lines had twists and bends, while others turned upside down or back on themselves. One man reported that he saw "sorts of woolly lumps at the tens" while another insisted that 9 was "a wonderful being of whom I feel almost afraid" and thought of 6 as "gentle and straightforward". A physicist described seeing numbers as in the form of a horseshoe, with 0 at the bottom right, 50 at the top and 100 at the bottom left. Another respondent, a barrister, described visualizing the numbers 1 through 12 as though on the face of a clock, with the following numbers tailing off afterwards into an undulating stream with the tens—20, 30, 40, and so on—at each bend.

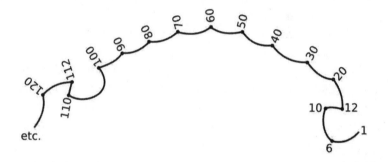

A recent experiment has confirmed the objective reality of such mental number lines. Remember the experiment that showed that, when asked to pick the larger of two numbers, subjects typically respond more slowly when the numbers are numerically closer together (53 and 55) than if the two given numbers are further apart (53 and 95). Researchers have found that, in most people who report a distinct mental number line, the response times for number comparison tasks do not vary with numerical distance, but with the distance in space between the numbers on their mental number line.

Some individuals also report using their mental number lines to help them perform arithmetic. One woman explained to researchers that wherever she went she was able to see groups of tens in lines one above the other. By moving around her number line she was able to calculate her change while shopping.

How I Calculate

All the evidence looked at so far in this chapter gives weight to my contention that the numerical abilities of savants such as myself are part of a broad spectrum of natural "number sense" that almost everyone has in one form or another. But to be sure, I need to consider alternative explanations for these abilities—two in particular posit a completely different mechanism from what we have seen so far.

The best known of these is what I call the "Sacks hypothesis," after the neurologist author Oliver Sacks, who popularized it in his writings. As described in chapter one, Sacks reports witnessing autistic twins instantaneously counting one hundred and eleven falling matchsticks—a scene later adapted for *Rain Man*, which replaced the matchsticks with toothpicks and increased the quantity immediately divined by Dustin Hoffman's character to 246. Sacks's account has also proven influential among a number of cognitive researchers. One theory put forward is that savants are able to visualize even very large numbers as quantities of individual dots in their mind's eye, which they then manipulate in their head to perform calculations.

Allan Snyder, director of the Centre for the Mind in Sydney, Australia, recently attempted to replicate "savantlike numerosity" under laboratory conditions. Snyder asked twelve healthy adult participants to count the number of dots briefly flashed up on a screen (between 50 and 150 at a time), following a short, painless application of magnetic energy to the left temporal lobe called repetitive transcranial magnetic stimulation (rTMS)—believed to temporarily replicate autisticlike, detail-oriented perception. In twenty trials, each lasting 1.5 seconds, both the spatial positions and quantity of dots were randomly selected by a computer. Snyder counted each trial as successful if the participant guessed the total to within five of the number of dots displayed. Before the TMS was applied, the subjects guessed correctly (within the margin of 5 either way) around 15 percent on average, compared to around 25 percent of the time following it.

The biggest problem with the theory that savants have an ability to instantly count very large quantities, which they use to perform calculations, is that no scientific experiment has ever demonstrated this ability, even more than twenty years after Sacks first wrote of it. Significantly, the subjects treated with TMS in Snyder's study could give only rough estimates, and then only a quarter of the time on average with any success— hardly analogous to the purported ability to instantly and precisely count 111 matchsticks. Savants who have been evaluated by scientists do not report being able to perceive and count large quantities in this way, myself included.

The second, even more speculative, alternative theory for savant numerical abilities, proposed by psychiatrist Diane Powell and computer scientist Ken Hennacy, is that they are somehow related to quantum mechanics. Powell and Hennacy believe that the current scientific models of the brain and for memory and learning cannot account for savant abilities. They argue that savants consider many possible solutions to a sum simultaneously in their minds to arrive at the correct result, something that can be happening only on a "quantum level" of consciousness. Powell and Hennacy's theory has been strongly criticized by other scientists, including Peter Slezak, a philosopher of the mind at the University of New South Wales. Slezak points out that there is no empirical evidence for the quantum consciousness theory and argues that savant numerical abilities are comparable in complexity to language abilities in most people:

> We're all savants in an interesting way . . . understanding language . . . there's an extraordinary level of mathematical complexity to the ability to do that which we

don't actually fully understand . . . we've evolved to do this automatically, instinctively, intuitively without effort. That's the sort of thing that savants do but that's just a different domain . . . you wouldn't resort to quantum physics to explain language even though . . . [it] invokes all sorts of equally complex forms of fast computations.

I agree with Slezak. In fact, my own theory for savant numerical abilities draws on this very analogy between the mathematical complexity of language and savant calculations. In order to describe this theory, I need first to explain a little about how brains function. In most people the major cognitive tasks—such as understanding language, figuring numbers, analyzing sensory perceptions, and so on—are highly specialized, performed separately in different regions of the brain. This specialization of different mental activities is effected by a process known as "inhibition," which prevents one area of the brain from interfering with the activity of another.

Various scientists have speculated on the possibility that a range of neurological conditions, from autism and epilepsy to schizophrenia, might be related to reduced levels of inhibition in the brain, causing abnormal cross-communication between usually separate brain regions. Cognitive scientist Ed Hubbard has argued that such activity within the brain might also account for the multisensory experience of synesthesia. Reduced levels of inhibition in the brain may also play a role in savant abilities: the savant Kim Peek was born without the corpus callosum— the thick band of tissues that connects the left and right hemispheres, and which also serves as the main inhibitory pathway in the brain.

I believe this abnormal communication between normally distinct areas of the brain is the starting point for an explanation of savant numerical abilities such as mine. It is very likely that my own brain works in this way—aside from high-functioning autism, I suffered from epileptic seizures as a young child and my father has long battled with schizophrenia, strongly implying a genetic cause for my brain's uncommon functioning. My ability to see numbers and language in different shapes and colors is an additional symptom of this unusual "cross-talk" between my different brain regions.

My hypothesis is that my numerical abilities are the result of abnormal cross-communication between the number and language regions of my brain. Specifically, I believe that my numerical abilities are linked to activity in the region of my brain responsible for syntactical organization (the forming of grammatical sentences in language).

Several pieces of evidence support this theory. First, the areas of the brain posited by researchers as being specialized for numbers (left parietal lobe) and for language (left frontal lobe) are next to each other in the left hemisphere. The left parietal lobe is involved in sequential and logical spatial abilities and appears to assist us in performing calculations, while the left frontal lobe—specifically Broca's area, is believed to house the ability to produce orderly, syntactic sentences. In essence, I am arguing that my ability to segment and manipulate number shapes in my mind to perform various calculations is analogous to the segmentation and manipulation of words and phrases into meaningful sentences.

A second piece of evidence comes from the extent of my linguistic ability: I know a dozen languages, can learn a new language to high conversational standard in a week, and am even creating my own language (more about this in the next chapter). It seems sensible to me to posit that some underlying mental activity links my linguistic abilities with my numerical ones, especially as both represent forms of complex computation. Given these facts, the alternative, that my linguistic and numerical abilities are unrelated, appears unlikely.

Third, my numerical abilities are rapid, intuitive, and largely unconscious—just like the syntactic computation in most people's minds when they are producing a sentence in written or spoken language. Over the next few pages I give a range of examples to help illustrate further the "languageness" of my numerical computations.

Let's start with a simple explanation of syntax as a system of rules that govern the positioning of various items (words) and their interrelations to one another. Syntax in language helps us to do more than just name things; it helps to describe and analyze how the various segments of speech interrelate. Syntactic constructions—whether in words or, as I argue in my case, numbers, too—make otherwise jumbled strings of information analyzable and meaningful.

The next thing to note is that words are immediately meaningful to most people, both because they relate to mental pictures and because words share semantic relations with other words. Think of the word "giraffe," for example, and you will likely immediately see a picture in your mind. You will also be able to think of other words that have something in common with a giraffe, such as "tall" or "neck" and so on. For most people, however, when they see or think of a number the same process does not occur. They do not see a mental picture as I do and as they do for words. Nor does a number generate other meaningfully related numbers, as the word "giraffe" naturally generates other related words like "neck" and "tall." If asked which numbers they think of when given the number 23, for example, many people will respond with the preceding number—22—or succeeding one—24—or perhaps by reversing it: 32. These are not especially meaningful (semantic) relations, but rather like being given the word "god" and replying with a word preceding it such as "go" or a word following it like "got," or by reversing it to produce "dog." In contrast, when I think of 23, I immediately think of meaningful (semantic) relations such as 529 (23 squared) or 989 (the last multiple of 23 before 1,000). I can do this because numbers do not exist in isolation but in meaningful relationships with other numbers—just as words occupy "semantic categories" with other words that help explain one another. Just as it would be impossible to talk about the word "giraffe" without words like "neck" or "tall," so it is impossible for me to talk meaningfully about the number 23 without referring to such relations as 529 or 989.

I know these semantic relations between numbers as I know the relationships of meaning between words, because I can visualize the numbers as meaningful (semantic) shapes. An English speaker given the word "chair" will be able to relate it to other words, like "stool" and "sofa," because he knows and understands the visual resemblance between them. In the same way, being able to visualize numbers helps me to see and understand the various interrelationships between them.

An important question here is where my number shapes come from—the simple answer is I do not know for sure. I do not know why I think of 6 as tiny and 9 as very large or why threes are round and fours pointy. There are patterns, however, showing again that my shapes are meaningful and nonrandom: 1 is bright, 11 is round and bright, 111 is round, bright, and lumpy, 1,111 is round, bright, and spinning. This suggests to me that my brain has taken a small number of synesthetic experiences for the smallest numbers and combined them in all kinds of ways to generate thousands of shapes—in the same way that languages take a small number of letters and sounds and generate thousands of words from them.

Of course, even the biggest of vocabularies cannot contain every word, and my "numerical vocabulary" of number shapes extends to 10,000 but not beyond. So how do I find the result of a sum when the solution is greater than 10,000—is, in other words, outside of my basic "vocabulary"? A useful analogy here comes from a recent game of Scrabble that I played with some friends. Midway through, I realized that I could play all my letters in one turn (winning fifty bonus points in one swoop), but I was not entirely sure whether the word I was thinking of— "agedness"—was in fact a word. The reason for this uncertainty is that "agedness" was not in my active English vocabulary, meaning that I had never used or read or heard it before—little wonder how rare it is; the "Wordcount" website, which lists the 86,000 most common words in the English language, does not have a listing for it. It *is* a word, though—WordNet, a lexical database created by Princeton University, gives the definition: "the property characteristic of old age." Fortunately for my score, I went with my intuition and played the tiles on the board and subsequently won the game.

How was I able to come up with a word that I had never heard of before? The answer is that though I did not know the word itself, I knew its components: the adjective "aged" and the suffix ending "-ness." Being a native English speaker, I was also intuitively aware of the morphological rules in English that allow certain adjectives—like "left-handed" or "broken" but not "strong" or "intelligent"—to end in that particular suffix. The same process of intuiting novel solutions from existing knowledge helps me to solve sums even when the result exceeds the limits of my numerical vocabulary. For example, given the sum 37 x 469, I immediately see that this is the same as 37 x 169 (something I can immediately access from my numerical shape vocabulary) plus 37 x 300 or 6,253 + (111 x 100) = 6,253 + 11,100 = 17,353.

This process of taking a sum and manipulating it in my head into meaningful number shapes and patterns that generate a solution is one I consider syntactic—analogous to how most people take a jumble of thoughts in their minds and effortlessly manipulate them mentally into a coherent, grammatical, and meaningful order that they can express as a sentence. For some reason they cannot do anything like this with numbers. Most people just seem overwhelmed by large numbers and are unable to think about them in the way they think about words in a sentence. In contrast, I am able to take the numbers in a sum (37 x 469) and mentally break them down into meaningful shapes (6,253 and 111 x 100), which I can then manipulate into a "sentence" that is grammatical (6,253 + 11,100 = 17,353) in the sense that it produces the correct answer.

Aside from multiplication, my other major numerical ability is for factorization and recognition of prime numbers. Factorization is the process by which a number is broken down into primes that, multiplied together, result in the original number: for example, 42 is factorized as 2 x 3 x 7. Being able to visualize numbers as semantic shapes helps me to factorize them rapidly. Using the example of 6,253 from the multiplication we just looked at, I can immediately "see" that its shape derives from the combination of 13 x 13 (169) x 37. This ability to instantly segment a semantic number shape is analogous to a native English speaker's to instantly segment a compound word like "incomprehensibly" as "in" + "comprehend" + "ible" + "ly." I factorize larger numbers (above 10,000) by separating them into meaningful (visualizable) related parts: for a number like 84,187, I separate it into 841 (29 x 29) and 87 (03 x 29), telling me immediately that the number is divisible by 29—as well as 2,903, a prime.

My ability to rapidly recognize primes up to 10,000 is a natural by-product of my highly structured, semantic numerical knowledge. My intuition tells me that 2,903 is probably prime, but to confirm this I mentally access the "region" of my numerical landscape for numbers between 2,900 and 3,000. Looking within, the first visually arresting numerical shape that I notice is 2,911—41 multiplied by 71. This tells me that the numbers before it (2,900–2,910) are all either easily divisible (by 2, 3, or 5) or prime. Unlike the "jaggedness" of composite numbers, I visualize the primes as smooth and round "pebbles" within my number landscape. I am able to generate larger primes, of 5, 6, 7, or even 8 digits, by drawing on my intuition for how prime numbers "look" in much the same way that a native English speaker can immediately tell that a word like "glubr" does not exist but one like "gluber" might. A very simple rule for perceiving a number's (non)primality would be: "If the number contains more than one digit and ends in 2 or 5, then it is not prime" (because it is obviously divisible by at least 2 or 5). A more complex rule would be: "If the number has four digits, the beginning and ending digit being odd (excepting 5) and identical, and the middle two digits being a multiple of 7, 11, or 13—such as 1,141 or 9,529, then the number is not prime" (because numbers of such a "shape" are always divisible by 7 or 11 or 13).

An example of how I can draw on intuitive rules like these to generate a larger prime is to start with a three-digit number that I know breaks down into small factors—like 323, which is 17 x 19—then to repeat it to produce a six-digit number—323,323—whose "shape" I recognize immediately as being also divisible by three other small factors (7, 11, and 13), and then to add a further digit which is not divisible by any of the factors so far listed, such as simply adding 1 to create: 3,233,231. This seven-digit number is indeed prime. Here is another example, this time using a four-digit number: 4,199 (13 x 17 x 19). We eliminate these factors by adding a single digit, 9, to the end of the number, to make 41,999. I add 9 because, separating the number into 419 and 99, I can see that it is not divisible by 11 (as 99 is, but 419—a prime—is not). I also know that 41,999 cannot be divisible by 7 because it is one less than 42,000 and 42 is divisible by 7. Separating into the number 41 and 999 I can also see that it is not divisible by 41 (as 999 is not divisible by 41) or 37 (999 is 3 x 3 x 3 x 37). 41,999 is prime. I can even repeat the process by adding another 9 to the end of 41,999 to make 419,999, which is also a prime number.

Of course, such an intuitive approach is not infallible, and I can make mistakes by thinking a number "looks" prime when it is not, or conversely when I think a number does not look prime enough and in fact is. Researchers have long known that savants sometimes make mistakes in identifying possible primes, contrary to the popular myth that our identifications are always flawless. For example, in her book, *Bright Splinters of the Mind*, psychologist Beate Hermelin describes a study of an autistic savant named Howard:

> When Howard was asked to find a prime number between 10,500 and 10,600, he came up with 10,511 in less than six seconds. He told the investigator that 10,511 was not divisible by 3 or 7. On further questioning . . . he replied, "13 will go into 611 and 10,511 minus 611 is 9,900, and that can't be divided by 13" . . . this condition [autism] appears to allow those who suffer from it a privileged access to segments and components of information . . . because of his autism he can use his cognitive style to allow him the fragmentation of target numbers into their components, which then enables him to apply possible divisors rapidly. On some occasions Howard simply said about certain numbers that they felt like primes. But his sense of intuition was not infallible and sometimes he was mistaken.

In fact, 10,511 is not prime, being divisible by 23 and 457, even though it "looks" like one. This fallibility in savants' ability to intuit prime numbers is analogous to a native English speaker's difficulty in using generally unconscious, intuitive word formation rules to distinguish some actual English words from nonwords. Here is a quick test to demonstrate this: simply identify which of the following words are actual English words and which are not:

Tsktsking
Syzygy
Ooecia
Gleever
Leiotrichous

Not easy, considering how many words there are in English. The same is true of primes: there are no fewer than 664,579 prime numbers up to seven digits in length. The answer: except for "gleever," which does "look" like it should be one, all are actual English words.

The Beauty of Mathematics

Any explanation of my numerical abilities would not be complete without an attempt to describe how much numbers mean to me, and, I am quite sure, to other savants, too. Social isolation and loneliness are common problems for individuals on the autism spectrum, as are feelings of frustration and confusion at a world that often seems too big, strange, and chaotic. Numbers can serve as a refuge—an interior universe of logic, order, and beauty. In this final section of the chapter, I want to share with you a little of the immense beauty I find in math.

Let's start with a simple problem. What is the largest number? Some children amuse themselves with this question, answering any proposed candidates with the two little words "plus one." There just cannot be a largest number, because numbers go on forever.

The idea of infinity has captivated mathematicians and philosophers for centuries. Galileo was the first to notice a remarkable fact related to the paradoxical nature of infinity: if you take the set of natural numbers (1, 2, 3, 4, 5, et cetera) and remove exactly half of them, the remainder is as large a set as it was before. For example, remove all the odd numbers and then compare the set of numbers left over (the even numbers) with the set of all natural numbers by pairing each of the numbers in the two sets: 1 with 2, 2 with 4, 3 with 6, 4 with 8, 5 with 10, and so on.

1,	2,	3,	4,	5,	...	n,	...
$\updownarrow$	$\updownarrow$	$\updownarrow$	$\updownarrow$	$\updownarrow$	...	$\updownarrow$	...
2,	4,	6,	8,	10,	...	2n,	

It is clear that we could continue pairing the numbers from both sets forever—meaning that both sets can be said to be of equal size. In fact, this is the very principle that mathematicians today use to define a set of numbers as infinite: a set is infinite if you can take out some of its numbers without reducing its size.

A beautiful illustration of the counterintuitive nature of infinity comes from the "Grand Hotel" paradox invented by the mathematician David Hilbert. Imagine a hotel with an infinite number of rooms. What happens when a newcomer wants to check in but all the rooms are already occupied? In fact, it is always possible to accommodate a new guest; simply move the guest occupying room 1 to room 2, the guest occupying room 2 to room 3 and so on, leaving room 1 vacant for the newcomer. What if an infinite number of newcomers arrive at the hotel? They, too, can be accommodated by moving the guest occupying room 1 to room 2, the guest occupying room 2 to room 4, the guest occupying room 3 to room 6 and so on, leaving all the odd-numbered rooms for these new guests.

There are an infinite number of primes, too—a fact first proved by the Greek mathematician Euclid more than two thousand years ago. Euclid's proof goes something like this: Let us start by assuming that there are a finite number of primes, which we write down—every single one—in a (very long) list. Next imagine multiplying all these prime numbers together, then adding 1. We now have a new number, P, which either is or is not prime. If P is prime, then it was missing from our original list of primes. If P is not prime, then it must be divisible by two or more prime numbers, except that it cannot be divided by any of the primes in our list without leaving a remainder of 1. But whatever prime it might be divisible by is also not on our list. Either way, the finite list was incomplete, meaning that there must be an infinite number of primes.

Prime numbers are fascinating and mysterious to savants and mathematicians alike: they seem to be randomly distributed along the number line, yet are capable of producing beautiful patterns. A good example of this is the Ulam spiral, named after the Polish mathematician Stanislaw Ulam, who discovered it while doodling to while away the time during a long lecture. Feeling bored, Ulam drew a grid of numbers starting with 1 in the middle and spiraling out. He then circled all the primes on the grid.

To his astonishment, the circled numbers appeared to form several long diagonal lines. Subsequent tests showed that the effect persists even when a very large number of numbers is plotted. Ulam's discovery was so unexpected that his spiral subsequently appeared on the cover of *Scientific American*.

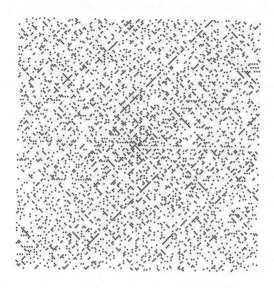

Primes are not only beautiful numbers; they are useful, too, in the field of cryptography. For example, whenever you make a purchase over the internet, the security of the information sent to other computers is ensured by the use of an encryption method known as RSA (the name derives from the initials of its inventors). Its efficacy depends on an extremely big number (e.g., 200 to 300 digits long) produced by multiplying together two very large prime numbers, both of which would need to be discovered in order to "unlock" the encrypted information. As the number of steps needed to factorize a number using all known algorithms increases exponentially with the number's size, the cryptographer can always stay one step ahead of the criminal's computer. Even if computers become fast enough to factorize the numbers currently used for enciphering, much larger numbers can always be generated to replace them.

We finish this briefest of surveys of the sheer beauty in mathematics with one of my favorite discoveries from a field known

EMBRACING THE WIDE SKY

as network theory, which examines how networks form and
function in society.

The "small-world" phenomenon, the theory's best known
concept, was an idea first written about by the Hungarian author
Frigyes Karinthy back in 1929. Karinthy imagined a scenario
where any two individuals could be connected through at most
five acquaintances. Four decades later, the psychologist Stanley
Milgram designed an actual experiment to count the number
of ties between any two people. Milgram chose a range of individuals from cities around the United States, such as Omaha,
Nebraska, and Wichita, Kansas, as the starting points, with Boston, Massachusetts, as the target city—selected because of the
great distance, socially as well as geographically, between them.
Information packets were sent to randomly selected individuals
in Omaha or Wichita with letters explaining the study's purpose
and basic information about a target person in Boston. If the
individual knew the target person himself, he was to forward
the letter directly to that person. In the more likely case that he
did not know the person, he was to think of a friend or acquaintance who might and forward the letter on to this person. The
letters were signed each time they were forwarded on so that,
if and when the letter reached its target, the researchers could
count the number of times it had been forwarded from person
to person.

Milgram's experiment suffered from a number of problems,
however, in particular that many people often refused to pass
on the letter. Among those letters that did make it to the target
person, the average number of forward-ons was between five
and six. Hence the researchers concluded that just six people on
average separate everyone from everyone else.

The math behind the idea is pretty simple. Let's assume that
a person knows on average 100 people, each of whom knows 50
other people, who in turn knows another 50 people and so on
up to six degrees. This works out to: 100 x 50 x 50 x 50 x 50 x 50

= 31,250,000,000 or over 31 billion. The current population of the world is a little less than 7 billion, making six degrees easily enough to cover its entire population.

A recent study by researchers at Columbia University has provided more evidence for the small world phenomenon. The "Small World Project" was carried out online, with each participant assigned a random target—1 of 18 people around the world. The participants were asked to link to their target by email via a chain of their friends and acquaintances. Some 60,000 people from 170 countries took part, and from the hundreds of chains completed, the researchers determined that the average number of links was indeed six.

Researchers have discovered that this "small world effect" can occur in any large network of linked, dynamic elements—from the national power grid and the internet to the brain and genome. Steven Strogatz, a professor of applied mathematics at Cornell University, is an expert of the phenomenon and has made important discoveries about what he calls the "universal architecture" of connection. Strogatz and his colleague Duncan Watts calculated that only a small number of shortcuts between a few components are necessary to produce the effect—linking clusters of people, websites, or brain cells together in unexpected ways. The scientists looked at a number of real-world systems to confirm their findings. In one study, Strogatz and Watts examined the nearly quarter of a million actors listed in the Internet Movie Database. The results confirmed their theory: a small number of well-known and prolific actors proved to be the hub of a highly structured network across which all kinds of actors could be linked to one another in just a handful of steps. For example, Alfred Hitchcock can be connected to Demi Moore in just three links: Hitchcock was in the film *Show Business At War* (1943) with Orson Welles; Welles starred in *A Safe Place* (1971) with Jack Nicholson; and Nicholson played a role in *A Few Good Men* alongside Demi Moore! In another study, the researchers found that any one of a roundworm's 282 neurons could be linked via their 2,462 synaptic connections within an average of 2 to 3 steps.

Little wonder then that so many people are left in awe by such results: each one of us is more closely connected to one another than we might ever have otherwise imagined. Our world is far smaller than most people realize—and far more numerical, too.

6

The Biology of Creativity

What is so different from the universal instincts for language and counting in my mind that enables it to create numerical landscapes (like those for Pi) from a random string of digits, invent my own words and concepts in numerous languages, and solve a sum by picturing it as the manipulation of complex shapes? The answer, I believe, lies at least in part in the unusual cross-communication between different regions of the brain that we looked at in the last chapter. Rare forms of creative imagination may be the result of an extraordinary convergence of normally disconnected thoughts, memories, feelings, and ideas. As I will show over the course of this chapter, such "hyperconnectivity" within the brain may well lie at the heart of all forms of exceptional creativity.

Leaps of the Imagination

Making something out of nothing—philosophers have long debated the mysterious nature and origins of these "leaps of the imagination." Plato believed that such leaps were the result of divine inspiration, a gift from the gods. The seventeenth-century French philosopher René Descartes disagreed, arguing instead that all creative thought was the product of the deductive mind rather than the Muses. The Enlightenment thinker Immanuel Kant, writing a century later, attempted a middle approach between the supernatural and rational explanations of creativity. According to his view, creation is a spontaneous, willful activity, independent of either divine assistance or prior rules.

Like the philosophers of the past, today's neuroscientists are seeking to understand what makes some people especially creative. Some believe the answer to this age-old enigma can be found in the biology of the brain. Synesthesia, a mingling of the senses, offers a valuable window on how the brain produces original, creative thoughts, according to neurologist V. S. Ramachandran, who has been studying the phenomenon for the past two decades.

Researchers make a distinction between "lower" synesthesia—where the synesthetic response is produced by an item's visual appearance alone—and "higher" synesthesia, where multisensory associations are formed from abstract concepts such as numerical sequences or quantities. Because abstract concepts like numbers are actually laid out in anatomical regions or "brain maps" inside our heads, it is this cross-activation between such regions which is so intriguing to scientists studying the possible biological roots of creativity.

Ramachandran argues that this ability to link seemingly unrelated concepts is central to the artistic process, citing studies that suggest that synesthesia occurs much more frequently among artists, poets, and novelists. All these people share a marked facility for metaphor and analogy, such as Shakespeare's "It is the east, and Juliet is the sun," in which he expresses Juliet's warmth and radiance by linking them to the sun's. The professor's theory proposes that such associations are facilitated by higher synesthesia in the artist's brain, allowing him to make novel connections between normally unconnected concepts much more easily and fluidly. Ramachandran suggests that this synesthetic capacity is genetic in origin, naturally predisposing an individual to highly inventive, original thoughts and ideas.

Other researchers have also examined the idea that some people are born more prone to certain forms of creativity than others. Nancy Andreasen, professor of psychiatry at the University of Iowa and a National Medal of Science winner, speculates on both the biological and environmental influences on highly creative individuals in her book, *The Creating Brain: The Neuroscience of Genius*. Andreasen cites a range of anecdotal evidence to support the idea that certain forms of creativity are inheritable, such as the variety of famous families in which at least two members made significant creative contributions, including the Darwins (Erasmus and his grandson Charles), the novelist Brontë sisters, and the Bachs, whose creative members extended over eight generations from 1550 to 1800.

Andreasen does not rely solely on anecdotal evidence, though, to support her hypothesis, performing her own scientific studies that put it to the test. In one such study, Andreasen compared the families of participants in a writers' workshop with those of a control group, dividing the two groups, relatives into three classifications: not creative, moderately creative, and highly creative. Examples of moderate creativity included occupations such as journalism, dance, or music teaching, while highly creative relatives were those who had written novels, performed as a concert artist, played in a major symphony, or made a major scientific contribution. The 30 writers had a total of 116 relatives, while the 30 controls had 121. Among the writers, 32 relatives (28 percent) were creative—20 very and 12 moderately—while among the controls 16 relatives (13 percent) were creative—11 very and 5 moderately. Mathematic analysis of the results showed that these differences were statistically significant, meaning that they were unlikely to be a chance result.

A plausible objection to Andreasen's study is that it does not tell us whether the disparity between the writers' and controls' relatives was the result of natural or environmental factors. Simply being born into a nurturing family environment might account equally well for the difference between the two groups. Andreasen acknowledges this counterargument, agreeing that, for example, literary people may teach their children writing skills at an early age and predispose them to become successful writers later in life. However, her study does provide some definite hints that the origins of artistic or scientific creativity are at least partially biological. Most significantly, the types of creativity found in the writers' relatives were not necessarily literary—many were in completely unrelated areas such as art, music, dance, mathematics, and science, suggesting that they could not have been the result of purely environmental factors.

A further perspective on the unique properties and biology of extraordinary creativity comes from the British mathematician and physicist Roger Penrose. In his book, *The Emperor's New Mind*, Penrose argues that sudden creative insights are decidedly uncomputerlike in their nature, rooted in complex processes deep within the human brain. A computer, Penrose argues, can be designed to process information within certain fixed parameters but is incapable of making the kinds of creative "leaps" exemplified by Archimedes' Eureka moment.

Penrose's reasoning goes something like this: all computers operate according to algorithms, rules that the computer follows step by step. However, sudden creative leaps of thought or imagination are not accomplished in this way. For example, mathematicians often come up with a new theory even before finishing the step-by-step calculations that formally demonstrate it. Penrose concludes from this that machines will never be developed that can think truly creatively like humans.

There is something to this argument, for creativity is not about following rules to reach a result, but rather about bending or even breaking the rules to create something truly original. A machine given the rules for painting a portrait could never come up with a Picasso, for example, nor, given the rules for music composition, could it produce a John Cage. This contention brings us back to my theory that extraordinary creativity is the result of hyperconnectivity—abnormal cross-activation between various regions of the brain—a phenomenon that provides a viable explanation for great creative leaps of the imagination. Hyperconnected brains are the very opposite of coolly calculating machines, operating not from any step-by-step mental rule book but instead as a kind of beautiful, swirling chaos that draws on information from all over the brain to arrive at results that are truly, breathtakingly creative.

If true, this idea suggests that certain neurological conditions, such as epilepsy or schizophrenia, in which hyperconnectivity is sometimes a significant feature, could enhance certain forms of creativity. Let us look at the evidence for this and what more it might tell us about the nature of highly creative thought.

The Storm Within

The poet John Dryden famously wrote in 1681 that "Great wits are sure to madness near allied, and thin partitions do their bounds divide," making the observation that genius and madness seem to go hand in hand. There are many well-known examples from history of great creative individuals who struggled throughout their lives with brain seizures or mental illness.

One of the most famous is the Dutch artist Vincent van Gogh, who suffered from temporal lobe epilepsy. Van Gogh described his seizures as "the storm within," which consisted of hallucinations, confusion, and floods of early memories. Following the famous ear-cutting incident (speculated as being the result of one of his seizures), Van Gogh wrote to his friend and fellow artist Paul Gauguin:

> In my mental or nervous fever, or madness—I am not too sure how to put it or what to call it—my thoughts sailed over many seas. I even dreamed of the phantom

Dutch ship and of *Le Horla* [a fictional work by Guy de Maupassant], and it seems that, while thinking what the woman rocking the cradle sang to rock the sailors to sleep, I, who on occasions cannot sing a note, came out with an old nursery tune, something I had tried to express in an arrangement of colors before I fell ill, because I don't know the music of Berlioz.

The journalist Eve LaPlante was among the first to identify the possible link between Van Gogh's temporal lobe epilepsy and his extraordinary creativity in her book, *Seized: Temporal Lobe Epilepsy as Medical, Historical, and Artistic Phenomenon.* LaPlante also cites other famously creative (and believed epileptic) individuals, such as: Lewis Carroll, Edgar Allan Poe, Gustave Flaubert, and Fyodor Dostoyevsky.

A further example of exceptional creative thought residing simultaneously in a mind plagued with illness is that of the Nobel Prize–winning mathematician John Forbes Nash, Jr., whose struggle with paranoid schizophrenia was made famous by Sylvia Nasar's 1998 biography, *A Beautiful Mind.* Nash, a brilliant mathematical thinker who made groundbreaking contributions to economics, suffered from severe bouts of schizophrenic confusion and paranoia over three decades, as a result of which he lost his family and spent long periods of time in mental hospitals. Happily, Nash made a remarkable recovery in his sixties and received a Nobel Prize in 1994 for his work in a field of mathematics known as game theory.

A 2003 study by psychologists from the University of Toronto and Harvard University provides a key piece of evidence for the biological link between the mental disorders and original creative thought of individuals such as Nash and van Gogh. Psychologists Jordan Peterson, Shelley Carson, and Daniel Higgins hypothesized that low levels of latent inhibition in the brain—which helps shut out stimuli the brain considers extraneous to its needs—might contribute not only to psychosis but also to original thinking, especially when combined with high intelligence.

Their theory was bolstered by tests of Harvard undergraduates that showed that those who had excelled in an area of creative achievement were seven times more likely to have low latent inhibition scores. The researchers conclude that low levels of latent inhibition may be beneficial when combined with high intelligence and good working memory, but not otherwise. As Peterson notes: "If you are open to new information, new ideas, you better be able to intelligently and carefully edit and choose. If you have fifty ideas, only two or three are likely to be good. You have to be able to discriminate or you'll get swamped."

The findings of this study dovetail neatly with the theory that hyperconnectivity can lead to heightened levels of creative thought and output. They also suggest a possible reason for the difference in life outcomes between my father, who has battled with schizophrenia for much of his life, and me. It is possible that my father's brain is similarly hyperconnective but unable to avoid being swamped by the sheer number of mental associations and chaotic flow of his thinking. Somehow I have been lucky enough to be able to control the creative "storms" within my head and, better still, to use them to produce a range of original creative contributions.

Providing further evidence for the proposed link between the brain's biology and rare forms of creativity are medical cases of individuals whose artistic creativity only emerged following neurological restructuring caused by trauma or illness. Tommy McHugh was a middle-aged builder in Liverpool, England, when a cerebral stroke left him confused and speaking in rhymes. Though never previously interested in art, McHugh became consumed by a passion to create, writing poetry, painting, and drawing in pencil and felt tip. He even drew large-scale murals on the walls of his house. More recently he has turned his hand to various forms of sculpture and carving. McHugh describes his mind as "like a volcano exploding with bubbles and each bubble contains a million other bubbles . . . bubbles of unstoppable creative ideas." His work has been exhibited at various galleries and won plaudits from prize-winning artists.

Alice Flaherty, a neurologist at Massachusetts General Hospital, understands McHugh's case better than most. After receiving one of the scores of letters he had written to neuroscientists around the world, hoping to better understand the source of his unusual creativity, Flaherty decided to fly to Britain to meet him. Her reasons were as personal as they were professional: she had herself been seized by an insatiable urge to write that lasted for several months, following a severe bout of postpartum depression—a transformation she subsequently described in her memoir *The Midnight Disease*. Flaherty theorizes that her boost in creativity was the result of changes in her brain's temporal lobe that stimulated her desire to write.

Unlike McHugh and Flaherty, Anne Adams was already an experienced creative by the time she showed the first symptoms of FTD (frontotemporal dementia), yet the originality of her artwork was augmented considerably by her illness. Originally trained in mathematics, chemistry, and biology, Dr. Adams had decided to change career track in her mid-forties after her son was involved in a near-fatal car accident. Having nursed him back to health, she set up an art studio, painting architectural portraits of houses in the West Vancouver, British Columbia, neighborhood where she and her family lived.

Because FTD is a progressive degenerative disorder of the brain, it was several years before Adams's symptoms became apparent. Her paintings, however, were already showing signs of the ongoing rewiring in her brain, becoming much more novel and varied. In one, Adams used patterned, intersecting red, blue, and yellow squares to render the visual experience of a migraine. In another, she painted hundreds of vertical figures to represent each of the bars of music in Ravel's *Bolero*, the height of each figure corresponding to volume, shape to note quality, and the color to pitch. Neuroscientists believe that the metamorphosis in Adams's art was the result of enhanced functioning in the right posterior brain, a part involved in integrating information from the different senses. This region's activity grew as her dementia increasingly affected the frontal brain areas that normally restrain it, unleashing torrents of creative ideas.

Such examples of extraordinary creativity are not always associated with illness. Linguists have long been intrigued by cases of healthy young children (usually twins) who invent their own languages that only they understand, without any special assistance or training from adults. The phenomenon—known as "idioglossia"—is rare but provides further insight into the contribution biology makes to rare forms of creative talent. Among the reasons that these children are driven to create their own words and language, I believe, is the fact that young children's brains are naturally "hyperconnected," the result of an excess of synaptic connections, which cause the brain to rapidly overdevelop in infancy. For example, by age three the infant has twice as many synaptic connections as he will have in adulthood. The young brain gradually prunes back many of these synapses as it develops, protecting it from information overload and allowing it to function more efficiently. Early childhood is thus a unique period of creative opportunity, the results of which—as we are about to see—can be truly spectacular.

Linguistic Big Bangs

"Dug-on, haus you dinikin, du-ah."
"Snup-aduh ah-wee die-dipana, dihabana."

No one but American twins Grace and Virginia Kennedy knew what they were saying to each other during this conversation at a California children's hospital, recorded by linguists in the 1970s in an attempt to decipher the girls' private language. Then age six, the girls had up till that time been mostly brought up by their elderly grandmother, a taciturn native German speaker who knew little English, and allowed them to spend much of their time by themselves conversing with one another. Without friends and having little contact with the outside world, the twins' exposure to their native language was analogous to that of the children of pidgin speakers. Like these children, Grace and Virginia were not content simply to mimic the fragmentary sentences they heard at home, creating their own language that contained many neologisms (new word forms) and novel forms of syntax (sentence construction)—signs of an innate creativity they had somehow unlocked.

Linguists in the language clinic at the San Diego children's hospital first discovered "Poto" and "Cabengo" (the names Grace and Virginia used for each other) after the twins were brought to the hospital by their parents, who were bewildered by their inability to understand them. The scientists, too, were initially perplexed by the fluency and seeming impenetrability of the girls' invented language. The clinic's director described it as being like "a tape recorder . . . turned on fast forward with an occasional intelligible word jumping out." Psycholinguists Richard Meier and Elissa Newport were brought in to decipher the unique tongue. They did this by slowing down videotape recordings of the girls' therapy sessions, phonetically transcribing the twins' dialogue in order to break down and analyze what they were saying. After studying more than one hundred hours of videotape, the linguists were finally able to communicate directly with the girls in their own language.

The results of this analysis and conversations with the girls showed that Grace and Virginia's language shared many similarities to what linguists call "twin speech"—mishmashes of two or more languages (where the twins are raised in a multilingual environment) or word malformations due to novel pronunciations of common words, such as pronouncing "school" as "bool." Many of the Kennedy twins' words, it turned out, were in fact mispronounced English ones, with some German influence, such as "nieps" for "knife" and "pintu" for "pencil."

Among these, however, the scientists also discovered words that they were unable to translate, such as "nunukid" and "pulana," that the girls had seemingly invented themselves. They had also created some thirty different words for "potato"—their favorite food. There were grammatical innovations in their speech, too, such as the use of the preposition "out" as a verb: "I out the pudatoo-ta" ("I throw out the potato salad").

Slowly, the twins began to learn English with the help of speech pathologists at the clinic. As their communication skills grew, their family decided to place them in separate schools in an attempt to stop them from using their private language. Unfortunately, the effects of their family's emotional neglect appear permanent; now in their thirties, Grace and Virginia remain developmentally disabled, working apart in menial jobs.

Even more remarkable than the case of "Poto and Cabengo," in which the twins created their own words and syntax within an existing language, is that of a group of deaf children in Nicaragua who created an entirely new language out of nothing more than ordinary gestures. Linguists have hailed the new language, Idioma de Signos Nicaragense (Nicaraguan Sign Language), as a "linguistic Big Bang." Nicaraguan Sign Language's origins go back to the 1980s. Before that time, the country's schools for deaf children were few, meaning that most had to make do with the gestures improvised within families. Then in 1981 a vocational school opened, allowing the children who attended to communicate with one another. This communication started to evolve rapidly as younger children, ages five and six, began attending. First, these children quickly acquired the crude signing of the older pupils. Then, somehow, they began to transform these signs themselves, until before long they were using their very own brand-new, sophisticated language.

Astonished by the children's self-generated language, their teachers invited scientists to come to their school in order to study it in detail. Judy Kegl, an American sign-language expert at Northeastern University, was among the first to dissect the children's complex hand sweeps. Kegl noticed that one of the things that distinguished the new "idioma" from simple gesturing was the trait of "discreteness"—where information is broken down into discrete parts. For example, in the expression "rolling down the hill," one word refers to the action (rolling) while another to the direction (down). The older children signed this expression with a single continuous movement. However, the younger children separated the movement and direction into different signs. The advantage of this segmentation of information is the flexibility it gives to the language, allowing the children to recombine the signs with others to form a wide range of meanings.

Most exciting for Kegl and her colleagues was the fact that the youngest children were creating their own grammatical features rather than simply extracting them from their parents' speech, as do the children of pidgin speakers. For example, just like the novel grammar created by the Kennedy twins, prepositions in Nicaraguan Sign Language are used as verbs, so a sentence like, "the book is on the table," would be signed as something like, "table book ons." The children's words can also possess surprisingly inventive forms: for example, the verb "to look for" is signed by brushing the back of the left hand repeatedly with the middle and ring fingers of the right. Others reflect their playful humor: their sign for "Fidel Castro" is a wagging finger combined with a V-sign close to the mouth.

This exuberant invention of original word forms by young children has been documented in a range of languages around the world and suggests that such creativity may be a natural part of the process by which some children acquire a full grasp of their native language. The Soviet psychologist Alexander Luria listed a number of neologisms by young Russian children in his book, *The Child and His Behavior*. Examples include: describing something that everyone uses as a "vsyekhny" (from the word "vsyekh," meaning "everybody's"); talking about a doll in a bathtub that has sunk but might float back to the surface as "vytonula" (literally "sank out" using a prefix to distinguish it from "utonula," which would simply mean it has sunk), and referring to a future job repairing sewing machines as being a "mashennik," from the noun "maschina."

Interestingly, studies show that children with autism are much more likely to create neologisms than their nonautistic peers. Indeed many children—and adults—with Asperger's enjoy inventing their own original word forms and puns. Examples from the scientific literature include: "paintlipster" (lipstick), "flappy" (a piece of paper or cardboard), "water bones" (ice cubes), and "pling" (pencil). Such inventiveness is, I would argue, further evidence in support of my theory that heightened creativity arises from hyperconnectivity in the brain, which is known to occur in autism.

A much more extreme linguistic form of this "autistic creativity" is "Mänti," a language I have gradually created since childhood, based on the lexical and grammatical structures of Baltic and Scandinavian languages (a particular fascination of mine) and containing many novel words, meanings, and concepts. Many of the words in Mänti are formed using analogies, so we have "hemme" (ant) from "hamma," (tooth) because ants are traditionally considered biting insects; "rupu" (bread) from the verb "rupe" (to rip, tear); "ausa" (to hear) from "auss" (ear); and "rodu" (face) from "rode" (to show). Mänti is also rich in compound words such as "vantool" (toilet, literally "water chair"); "päivelõer" (diary, literally "day book"); "lugusopa" (shampoo, literally "hair soap") and "melsümmi" (bee, literally "honey fly"). Merging two words into one is another technique I use to create new words: "puhe" (to speak) and "kello" (bell) to make "pullo" (phone). Some compounds are used to express novel concepts such as "kellokült" for lateness or tardiness (literally "clock debt"). Numerous words in Mänti are onomatopoeic, especially those for the names of animals: "karka" (crow), "huhu" (owl), and "mää" (goat) are examples.

Many of the language's features are very different from those found in most European languages. For example, words for pairs (e.g., eyes, ears) in Mänti are always treated as a whole, so to say "one eye" you would say "puse aku" (literally "half an eye," meaning half the pair). Uncountable nouns (nouns that cannot be pluralized, such as "furniture" and "information") are expressed using two related countable nouns in the plural: "lenta toolt" (literally "tables chairs") and "sot kupat" (literally "words pictures").

Another typically Mänti characteristic is that periods of time are expressed by analogy to the typical duration of common activities, so "rupuaigu" (literally "bread time") would mean "about an hour"—the time it takes for bread to bake in the oven. Another example is "piippuaigu" (literally "pipe time"), a shorter period than the previous example, equivalent to the time it takes to smoke a pipe.

Mänti's grammar is just as exotic (for English speakers). For example, it is possible to generalize a word by repeating it, adding an "m" to the second word, so "armo" (love) becomes: "armo marmo" ("love, affection, devotion"). When the word already starts with an "m," a "v" is added: "meri" (sea) becomes "meri veri" ("a vast body of water"). Repetition is also used to intensify an action or description. Repeating a verb makes the action longer, so "lue" (to read) is different from "luelue" (to read for a long time). Similarly, repeating an adjective is equivalent to using the word "very" in English: "löbö" (good) and "löbö löbö" (very good). Repeating a noun indicates genuineness: Pinocchio, for example, wanted to become a "poipoig" (a real boy).

Are such examples evidence of a special relationship between autism and certain forms of creativity? The positing of such a link is not new: "It seems that for success in science and art," wrote Hans Asperger, the Austrian doctor who pioneered the study of autism in the 1940s, "a dash of autism is essential." Yet the possibility that great creativity might be found in the autism spectrum appears to challenge the standard definitions for both. In the sweetest of ironies, the traditional scientific understanding of creativity is being transformed by the discovery of the innovativeness of the autistic mind.

Autistic Creativity

Scientists once considered autistic thought to be the antithesis of creativity: learning-disabled, rigid, overly literal. Even the capacities of autistic savants were considered to be little more than an acute form of mimicry or obsessiveness. Such notions have been upended in recent years by a range of studies showing that individuals with autism are capable of considerable creativity and enrich our understanding of what it means to be truly creative.

First a note of explanation: The scientists' dismissal of possible creativity in autistic individuals was based on a misunderstanding: the standard criteria for diagnosing autism in a person include "poor flexibility in language expression and a relative lack of creativity and fantasy in thought processes." But researchers who began to recognize high-functioning forms of autism, such as Asperger syndrome, in the 1990s realized that these cases did not fit this, and other aspects of the old criteria. It quickly became clear that the supposed lack of creativity within autism was really the result of how it had previously been defined, rather than being a reflection of how those with autism truly think.

A further reason for the scientists' error is the difficulty in evaluating something as mercurial and enigmatic as creativity. In the absence of any simple, agreed-upon definitions, autism researchers have generally relied on traditional, easy-to-administer tests such as the Torrance Test of Creative Thinking (named after its creator, educational psychologist E. Paul Torrance). However, such tests bear an uncanny resemblance to the weaknesses of IQ tests for intelligence. For example, in a 1999 study by Cambridge University's Autism Research Centre, researchers using the Torrance test showed a group of autistic and nonautistic children a toy elephant and asked them to think of as many ways as they could to make the toy "more fun." The children with autism came up with fewer responses, prompting the researchers to conclude that they were less creative. Such a conclusion, though, appears to be a product of the test's sheer banality. After all, the concept of "fun" is vague and subjective. Confusion, rather than any lack of creativity, is a likelier explanation for the study's findings.

When allowed to express themselves through a medium that actually interests and engages their imagination, high-functioning autistic individuals do demonstrate considerable creativity, and some have even established careers using their creative gifts. One such is George Widener, a Cincinnati native in his forties, who has combined his fascinations for magic squares and calendars to create original art. Widener's squares use key dates from the lives of historical figures, such as Queen Victoria, to produce a "calendrical portrait" of the individual. His work has been praised by magic square and art experts alike, and has been displayed in exhibitions around the world.

High-functioning autistic individuals can also excel through words as well as images. In recent years, researchers have documented a remarkably wide range of original poetry by individuals on the autism spectrum. This should come as no surprise when we consider that certain figures of poetic speech, such as metonymy—the substitution of a word or idea with one related to the original (such as "crown" to mean "king") and metaphor, are comparable to the richly associative style of thinking common to the autistic mind.

For example, Temple Grandin, a professor of animal science and author who has high-functioning autism, describes her thinking as associative links based on mental images. Her thinking might jump from bicycles to dogs because she has seen dogs chasing bicycles. We find even more striking metonymic associations in Clara Claiborne Park's book, *Exiting Nirvana*, wherein she describes her autistic daughter Jessy's at-times highly poetic perceptions of the world around her. For Jessy, the number eight is "good" and "silence," while seven is between silence and sound, three is "doing something fairly bad," two is "bad," and one is "very bad." Happiness, to her, is "zero clouds and four doors."

Though not all autistic individuals are able to use or understand metaphors, many do perceive tangential similarities between different things, helping them to evoke complex emotions or ideas. University of London psychologists Beate Hermelin and Linda Pring analyzed several poems by an autistic woman named Kate, and found numerous examples of metaphors in her work. In one of her poems, for example, Kate describes herself as a "puzzled jigsaw" and in another as "a something where fog lingers somewhere." Like Kate, I also write poems to explore and express emotional states and experiences that are special to me. In the poem below, written during a trip to Iceland in May 2007, I describe my deep feelings of affection for this little nation and its people:

> Yesterday I went to Gullfoss
> Appeared a rainbow there
> I stepped on it by mistake
> And climbed into the sky
>
> Looking down I could see
> The light-swept land
> Wet moss and gleaming stones
> Bathed in warm and rippling air
>
> I saw my friends, like angels
> Disappear into the shining spray
> Wearing the waterfall
> Close against their skin, against their hearts

Elsewhere I saw rivers, their floors coated
With travelers' silvered hopes
Flung below like falling stars
Into the streaming darkness

In the distance I could see
Turrets of steam
Pulling at the horizon

And in the towns and cities
I watched people talking among themselves
Stitching their breath
With soft and colored words

In a harbor "sólfarið,"
A sunfaring man
With outstretched arms
Hugs time
Remembering the tide-washed dreams of men
Born and those still yet to be.

This fluent ability to form associations between seemingly unconnected objects or ideas appears to be at the heart of autistic thought and extraordinary creativity (both artistic and scientific)—an observation that goes a long way towards explaining some of the peculiar features of both. A famous example of scientific creativity is the discovery in 1865 of the structure of the benzene molecule by the German chemist Friedrich Kekulé—an event said to have ushered in the science of organic chemistry. Kekulé's insight—that benzene had a closed, hexagonal structure—came from a daydream in which he saw a snake biting its own tail. Kekulé awoke "as if by a flash of lightning," and spent the night developing his hypothesis, which he subsequently presented in a formal paper to the Royal Academy of Belgium.

Kekulé's great imaginative insight is just one among many examples of the creative leaps achieved by some of history's best-known scientists. Biographical analyses of several of these—including Isaac Newton, Albert Einstein, Nikola Tesla, and Gregor Johann Mendel—by Michael Fitzgerald, a professor of psychiatry at Trinity College Dublin, suggest that they may all have found their genius through autism. Mendel, an Austrian monk and botanist whose discoveries laid the foundations for the modern science of genetics, had an extraordinary passion for counting all kinds of things: peas, weather readings, students in his classes, and bottles of wine purchased for the monastery cellar. Fitzgerald notes that for his experiments with peas, Mendel would have counted a total of more than 10,000 plants, 40,000 blossoms, and 300,000 peas. He concludes: "Virtually no one except a person with autism could do this."

Fitzgerald believes the creative achievements of the scientists he lists were the fruit of particular features characteristic of Asperger syndrome, including intense focus on a topic, great powers of persistence and of observation, enormous curiosity, and a compulsion to make sense of the world. Asperger talents, Fitzgerald contends, "changed the world."

The neurological mystery of creativity may never be solved, and maybe that is for the best. But the idea that great creative acts are rooted in our biology does help to explain the persistence throughout our evolution of conditions such as epilepsy, schizophrenia, and autism. Even more, it reminds us of the humanity of even the world's greatest creative geniuses. The music of a Mozart or the paintings of a Picasso have the capacity to touch each person profoundly, because they were made from the same fabric as that of every other mind. As Shakespeare noted, we are each the stuff of dreams. The simple reason then that great creative works enrich us is that they serve as reminders of the treasures buried deep within us all.

7

Light to Sight

The French novelist Anaïs Nin famously observed: "We do not see things as they are, we see things as we are," suggesting a vision of vision that is as personal as it is biological. Far more than simply a matter of focusing our eyes, our perceptions are patchworks of reflected light, emotion, and expectation. They remain puzzles, too, for researchers who hope that a science of seeing will yield new insights into the workings of the human mind. Our sight is a marvel of engineering, considering its complexity and its diversity, for, as Nin and the neuroscientists agree, no two people see alike.

How I see the world illustrates the uniqueness and individuality of vision. My autism gives me a perception that is highly detail-oriented. For this reason, one of my favorite childhood books was *Where's Waldo*, which requires a good eye for detail. To this day I regularly spot misspellings and other subtle errors in the pages of a book or newspaper. When I step into a room for the first time, I often feel a kind of dizziness with all the bits of information that my brain perceives swimming inside my head. Details precede their objects: I see the scratches on a table's surface before seeing the entire table; the reflection of light on a window before I perceive the whole window; the patterns on a carpet before the whole carpet comes into view.

An explanation for my piecemeal perception comes from autism researcher Uta Frith's "weak central coherence" theory. Central coherence describes the ability to pull together large amounts of information into a meaningful whole. According to Frith, this capacity to synthesize bits of information appears to be altered in autism, causing a focus on detail at the expense of gist, or the "bigger picture."

Scientific studies appear to bear out her theory. In the Navon task, individuals are presented with a series of letters composed of smaller ones and asked to press a left or right button for each, depending on whether they see a target letter or not. For example, a subject might be presented with a large *A* comprised of smaller *H*'s and subsequently asked whether he saw an *A* or not. Researchers at Cambridge University's Autism Research Centre gave me the task and found that my ability to perceive the large letters was impaired by my instant perception of the smaller ones. In most people, the reverse is usually true.

Cognitive psychologist Francesca Happé has shown that this detail-focused style of visual processing has its advantages. In the Ebbinghaus illusion, two identically sized circles are presented side by side. One circle is surrounded by a ring of large circles, and the other by a ring of small ones. Happé showed this optical illusion to a group of autistic and nonautistic subjects. She found that, while many nonautistic individuals were fooled into seeing the left circle (below) as being larger than the right, the viewers with autism were far less likely to be tricked in this way.

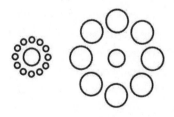

Here is another example of the difference between autistic and nonautistic perception. Once, while browsing through a book in a bookstore, I came across the following puzzle and asked my (nonautistic) friend beside me to give it a try: How many *F*'s are there in the sentence: "Finished files are the result of years of scientific study combined with the experience of many years." My friend carefully read the sentence over and answered "three." In fact, as I showed him, there are six. The reason most people fail to spot the other *F*'s is that they only look for them in nouns, overlooking those in the preposition "of."

Each person's perceptions are filtered by his assumptions about what he is seeing. This is why my friend thought he saw three *F*'s instead of six. Even though his eyes could see all of them, his brain was only searching for information where he was expecting to find it.

The selectivity of people's perceptions can sometimes cause them to miss even seemingly conspicuous details, as demonstrated in an amusing experiment by psychologists Daniel Simons and Christopher Chabris. The scientists played a video of people playing basketball to a group of subjects, asking them to keep count of the number of passes made by one of the two teams. So focused were the volunteers on their task that around half failed to notice a person dressed in a gorilla suit walk slowly among the players, face the camera, and thump his chest, before exiting the other side.

Look Smarts

For all these incongruities, our "eye-q" makes visual virtuosos out of us all. Without even having to think, our brains absorb huge quantities of information—line, color, shape, shading, depth, motion, and more—and all in the blink of an eye. Just as impressive is the fact that what we see is the result of thousands of continuous tiny adjustments by our eyes and brain, providing us with images that appear seamless and are obtained effortlessly. So rapidly does this automatic updating happen that any glitch is almost never perceived.

So how exactly do we see? Aristotle thought that observed objects altered the "medium" [air] between themselves and the individual, a change that was then somehow picked up by the eye. This passive view of perception, or variations of it, remained the standard explanation for centuries. Not until recent times did scientists discover that vision is in fact a far more active, constructive process than had ever previously been imagined.

Vision starts with light rays bouncing off the surface of a viewed object and entering the eyes. The pupil (in the center of the iris) determines how much light to let in, getting bigger in dimly lit situations and shrinking when in bright light. After passing through the cornea and pupil, and then through the lens, the light hits the retina, a soft, light-sensitive layer of nerve tissue that lines two-thirds of the back of the eye. We actually see with the retina—sight would be impossible without it, no matter how much light passed through the eye.

The retina is composed of two types of specialized cells (known as photoreceptors) called rods and cones. There are around 120 million rods in each eye, helping us to see in dim light and to differentiate shades of gray. In comparison, the six million cones allow us to figure out how much red, green, and blue is in each part of what we are looking at. By mixing these color signals together in various ways, we are able to see every other kind of color.

When the reflected light hits the retina, its cells collect and turn it into electrical impulses. These are then sent via the optic nerve to various parts of the brain's visual center, which respond to different aspects of the perceived object, instantaneously fitting them together like a jigsaw puzzle to create the final image that the viewer sees and understands.

That we construct a comprehensive and comprehensible visual representation of our environment from the minute smudges of reflected light on our retinas is one of the truly awe-inspiring discoveries of modern science. However, it also remains one of its profoundest mysteries. Consider for example the fact that the image cast at the back of the eye could be interpreted in a potentially infinite number of different ways. After all, the retinal image is two-dimensional (height and width) but is converted by the brain into three (height, width, and depth). How does it choose between the countless possible three-dimensional interpretations for each image it receives? The eighteenth-century philosopher George Berkeley recognized this problem in his New Theory of Vision: "It is, I think, agreed by all that distance, of itself and immediately, cannot be seen. For distance being a line directed endwise to the eye, it projects only one point in the fund of the eye, which point remains invariably the same, whether the distance be longer or shorter."

In other words, the rays of light that reach the eye at any given moment could theoretically have reflected off an object inches, miles, or even light-years away. It is simply impossible to obtain such information from the retinal image alone. The ability to translate such ambiguous data into meaningful visual representations must then involve a series of unconscious rules that allow us to construct our visual worlds. Over many years of careful study, researchers have discovered dozens of these rules, and believe that they exist naturally in everyone.

The theory of a universal, innate "grammar of vision" is supported by research showing that babies react to motion; construct the boundaries, shape, and depth of objects; and use shading and perspective in their first year of life. No parent teaches his child how to see, because most do not understand how sight works, themselves. Instead, the individual's visual world appears naturally and spontaneously during early childhood, in much the same way as his native language does.

As cognitive scientist Donald D. Hoffman points out in his book, *Visual Intelligence: How We Create What We See*, an important benefit of such innate rules is that they lead to consensus in the visual constructions that most adults make. Given a new image, two people from opposite ends of the world will see, mostly, the same scene. Despite potentially wide cultural differences, their shared cognitive structures guide their visual constructions to similar conclusions.

This process of constructing what we see happens step by step, with one stage typically depending on the results of constructions at other stages. Hoffman gives the example of a book's shape in three dimensions being constructed from the results of constructing its motion, lines, and vertices in two dimensions.

A scene's depth and color are similarly subject to these rule-guided constructive processes. Perceiving depth, for example, is achieved by a mechanism known as stereoscopic vision. When we look at an object, the horizontal distance between the two eyes creates small differences in the images captured by both, which see the same scene from slightly different angles. The brain instantly fuses the different perspectives together into a single image, giving us the impression of depth.

Our minds are able to perceive the redness of roses or the blueness of violets with the help of photons, "particles" of sunlight. There are actually several different kinds of photons, corresponding to the different colors, with each having a different wavelength (violet being the shortest and red the longest). When a stream of photons falls on the surface of an object, some of the wavelengths are absorbed while others are reflected. Our minds construct the object's color from the distribution of the wavelengths that enter the eye.

The cognitive process of perceiving color is part of a larger one in which the brain constructs several visual properties at once, attempting to make them mutually consistent. So at the same time the brain is constructing an object's color, it is also giving it a three-dimensional shape and light sources (usually overhead) that illuminate it.

Martian Colors and Ink Blots

Color is a good way to illustrate the remarkable variety and subjectivity of our perceptions. The range of colors that we see when we look around us, though impressive, is in fact far from exhaustive. Humans are normally able to perceive colors with wavelengths of 400 nanometers (violet) through to 700 nanometers (red), with purples, blues, greens, yellows, and oranges in between. Birds, however, can see what we cannot —ultraviolet-range colors with shorter wavelengths (between 340 and 400 nanometers). For every color that we see, our feathered friends see many more.

For example, when humans look at a pair of *Parus caeruleus* (European blue tits), the male and his mate both appear alike, with an identical blue patch on top of the birds' heads. From a bird's-eye view, however, the male is quite distinct from the female. His "blue" is in fact an entirely different color—an ultraviolet-enhanced blue that is invisible to the human eye.

Some people also see the world of colors differently. An obvious example of this is color blindness, a condition caused by the inheritance of a faulty color vision gene. Color-blind individuals typically confuse red and green, whose different shades all appear dull and indistinct. Though not technically considered a form of color blindness, age-related changes to the cornea can also affect the ability to perceive violet and blue colors.

A remarkable example of how our biology can affect perception of color is seen in the case of a color-blind synesthete. Though the man's retina distinguishes only an extremely narrow range of light wavelengths when looking at objects, his brain's color area works just fine. Having a numerical form of synesthesia, he "sees" numbers in his mind's eye as tinged with hues he cannot otherwise perceive in the real world, referring to these as "Martian colors" that seem to him "weird" and "unreal."

The language we use to talk about colors can also affect our perception of them. For example, Russian speakers use two distinct words to describe an object's blueness: "siniy" (dark blue) and "goluboy" (light blue). In a 2007 study, MIT researcher Jonathan Winawer and colleagues devised an experiment to evaluate whether this linguistic distinction gave Russian speakers an edge over English ones. Winawer and his team recruited a group of fifty subjects, half of whom were native Russian speakers. The volunteers were presented with groups of squares in different shades of blue and asked to indicate which two in each set were exactly alike. The results showed that the Russian speakers were indeed faster at distinguishing between light and dark shades of blue.

Studies such as Winawer's confirm the influence of environmental as well as biological factors in our perception. The role of context in how a person perceives something helps to illustrate this point. When I type "laughter" and "12345678," your brain perceives the first character in the first example as a letter, and in the second as a number, based on their respective contexts.

Context plays an important role in art. For example, beginners are often encouraged to draw an object by concentrating on the space around it (known as "negative space") to help them render it more accurately. The idea is to encourage the artists to observe carefully what is before them and so prevent assumptions from leaking into, and subtly altering, their perceptions.

Some contexts can even make us "see" things that are not there. In a 2008 study, researchers at University College London were able to trick a group of subjects into "seeing" a nonexistent rectangle on a computer screen. They did so by presenting the volunteers with a series of vertical rows of small dimly lit rectangles. While many of the rows were complete, others had a narrow gap where the central rectangle had been removed. The viewers were often unable to perceive these gaps because their expectations unconsciously filled them with an imaginary rectangle that completed the line.

Magicians regularly make use of the effect past experience and expectations have on an observer's perception to produce various illusions, such as the disappearing ball trick. The magician throws a ball into the air, followed by a second and then a third, which magically "disappears." In fact, there is no third ball. The context, including such cues as the magician's upward gaze, fools the spectators' brains into thinking that something disappeared.

Visual perception can also be manipulated by our other senses. In one study, observers were asked to judge how many white dot flashes were displayed on a black screen. The flashes were accompanied by a number of auditory beeps, sometimes identical to the number of flashes displayed but other times different. The scientists found that a single flash shown with more than one beep was often incorrectly perceived as being repeated, too.

Most intriguing is the link scientists have discovered between hand position and visual processing. Washington University cognitive psychologist Richard A. Abrams and colleagues examined the possibility that an object's proximity to the viewer's hands might affect how well he sees it. The team asked volunteers to identify letters on a computer screen over two sessions. In the first, the viewers were asked to place their hands beside the monitor as they carried out the task, while in the second they put their hands in their laps. The scientists found that the volunteers performed better when their hands were closer to the screen, even when they were hidden behind cardboard. Abrams speculates that the brain's visual center gives special attention to the area around our hands because of their role in a range of crucial tasks, such as eating food and holding objects.

Our perceptions are most variable when the item we are looking at is ambiguous. A classic example of this is how a cloud resembles a wide range of objects, depending on which observer you ask. Inspired by the human tendency to project different interpretations onto ambiguous stimuli, the Swiss psychiatrist Hermann Rorschach created his famous inkblot test in the 1920s. Once dubbed an "X-ray of the mind," the Rorschach test uses ten symmetrical (half colored and half black-and-gray) ink-blots which the examiner asks the viewer to interpret. According to its practitioners, a person's responses reveal aspects of his emotional and mental states. Many scientists, however, criticize the test as being too subjective and unreliable.

Sleights of Mind

Ambiguous images have long been a staple of optical illusions—images that are designed to demonstrate the vagaries of our perceptions. You can see the the German psychologist Ernst Mach's "open-book" illusion below, for example, as opening either towards or away from the viewer. Once the viewer sees both perspectives, his eye will oscillate between the two representations as the brain tries to make sense of what it is seeing.

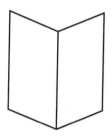

Among the many other kinds of optical illusions, geometrical illusions are the most common, involving distortions in our sense of length, height, or distance. Mario Ponzo, an Italian psychologist, hypothesized that the human mind gauges an object's size based on its background. He illustrated his argument with the following diagram, where the line in the distance appears longer than the line nearer the viewer, even though the two lines are in fact identical:

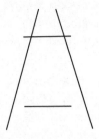

The Müller-Lyer illusion is another famous geometrical illusion, where the lengths of two identical lines are distorted by the angles attached to the ends; the more obtuse the angle, the longer the lines appear.

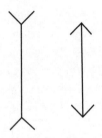

Below is a form of perceptual illusion ("the Penrose triangle") known as an "impossible object," where a two-dimensional image is interpreted by the brain's visual system as three-dimensional, even though such an object could not actually exist. The triangle is named after British psychiatrist Lionel Penrose and his mathematician son Roger who popularized it in the 1950s.

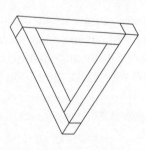

The popularly-named "devil's tuning fork" illusion is a further example of an impossible object, appearing to have three cylindrical prongs at one end but only two rectangular ones at the other.

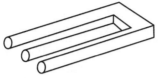

Scientists have long been fascinated with perceptual illusions and what they might teach us about how the brain processes vision. Neurobiologist Dale Purves describes his theory of human perception, based on his research using perceptual illusions, in his book, *Perceiving Geometry: Geometric Illusions Explained by Natural Scene Statistics.*

Purves and his colleagues employed a range of illusions to help them understand the brain's strategy for perceiving the world around it. In one study, the scientists found that viewers perceive an object as longer when it produces vertical or leaning lines in the retinal image, and shorter when it generates horizontal ones. They argue that this is because the possible real-world sources of vertical or near-vertical lines are, on average, physically longer than for horizontal ones. According to this theory, the view we see at any given moment is what our brain considers the most likely interpretation of the retinal image, based on previous visual experience. Optical illusions happen when the brain guesses wrong, selecting an interpretation that in fact differs from the retinal image's actual source.

Another view comes from cognitive scientist Mark Changizi, who argues that many perceptual illusions are the result of how the brain processes motion by anticipating what it is about to see. It takes the brain at least one tenth of a second to construct a visual model of what it is seeing, a time lag that means it is working with old information. The brain makes up for this by guessing ahead in order to "see" the present. An experiment by one of Changizi's colleagues, Romi Nijhawan, gives support to this theory. Subjects were shown an object moving past a flashbulb that flashed precisely at the moment the object was passing it. The observers, however, reported perceiving the object as having passed the bulb before it flashed.

The scientists use a sports analogy to help illustrate this further. When a baseball player swings his bat, he does not have time to wait for his brain to construct a visual representation of where the ball is in relation to him. Instead, he aims at where his brain predicts the ball will (very shortly) arrive.

Not all illusions are the creation of vision researchers—others are the result of the variety and complexity of the natural world. Camouflage, for example, allows an otherwise visible animal to become indiscernible from its surroundings, helping it to avoid predators and catch prey. Cryptic coloration is the most common form of camouflage, where the color of an animal matches its environment. Examples include the earthy tones of squirrels and deer and the white fur of the polar bear.

The moon illusion offers another example of how Mother Nature can fool our eyes. When the moon is at the horizon, it appears much larger than when it is high in the sky. Explanations for the effect go back to ancient times. Ptolemy and others thought that the atmosphere between the observer and the moon somehow inflated its size from the observer's point of view, but we have known for several centuries that the image of the moon on the eye's retina is constant whether it is seen on the horizon or not. Modern theories focus on physiological or psychological explanations for the phenomenon. One suggests that the lack of distance cues in the night sky causes the eyes to perceive the high moon as smaller.

As we saw earlier, magicians are, like Mother Nature, experts in the use of perceptual illusions that confound spectators' expectations. A remarkable example of this expertise comes from the story of British magician Jasper Maskelyne, who quit the stage at the start of World War Two determined to put his training as an illusionist to good use on the battlefield. As told in his memoir, *Magic: Top Secret*, Maskelyne was posted to Africa, where he joined a counterintelligence unit known as the "A Force." Charged with the task of concealing British forces from German aerial reconnaissance, the illusionist assembled a team of men with backgrounds in chemistry, engineering, and stage set construction to help him develop illusions that would perplex the enemy. The group, informally titled the "Magic Gang," built dummy tanks made of plywood and painted canvas—even devising a method for faking tank tracks in the sand after the dummies had been placed in position.

The gang's most spectacular illusion was created in 1941 when Maskelyne and his colleagues helped the major port of Alexandria to temporarily "disappear." The effect was achieved by creating a life-size replica of the port nearby, complete with dummy buildings, a dummy lighthouse, and even dummy anti-aircraft batteries that fired thunder flashes.

When not being used to create scientific theories of perception, camouflage animals, or divert enemy bombers, optical illusions appear regularly in artwork aimed at generating a range of perceptual responses in viewers. This is particularly true of "op (optical) art," a school of geometric and abstract art that first emerged in the mid-1960s. Op art paintings evoke numerous perceptual illusion effects arising from the precise and systematic manipulation of shapes and colors by the artist. Victor Vasarely, the movement's leading figure, used Necker cubes—which appear to protrude from and into the picture simultaneously—in many of his works.

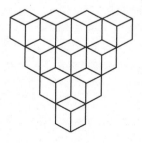

Famous also for his integration of various perceptual illusions (including Necker cubes) in his work was the Dutch artist M. C. Escher. In *Ascending and Descending*, a lithograph print made in 1960, Escher depicts a large building with the illusion of a never-ending staircase. He implemented a similar illusion using the Penrose triangle in his 1961 print *Waterfall*, which shows a waterfall which appears to flow up and down, then up again in an infinite loop.

The most common artistic illusion, however, is found not in museums or galleries, but in the cinema. Whenever we see a movie, the moving images we perceive as a continuous sequence are in fact a series of separate pictures (frames) flashed on screen at a rate of 24 per second. This rapidity allows each frame to linger on the viewer's retina just long enough to merge with the next.

The Psychology of Art

Many thinkers have pondered the relationship between art and perception and what each might tell us about the other. The influential art historian Ernst Hans Gombrich extensively wrote about how the principles of perception might produce a greater understanding of the creation and appreciation of art. One of Gombrich's most important claims was that painters do not simply paint what they see, but what they have learned to see. The observer's role is considered similarly interpretive; when looking at an artwork, he must use his mind to "collaborate with the artist . . . to transform a piece of colored canvas into a likeness of the visible world."

Underlying the whole of Gombrich's approach was his view that the miracle of art was not its ability to simulate reality, but to train the observer to see the world afresh. The artwork invites the individual not only to look outside but also within, to the subjective memories, ideas, and emotions that form his interpretation. This capacity of art to help us look within our own minds has captured the interest of some of the world's leading neuroscientists.

V. S. Ramachandran and William Hirstein caused something of a stir in the art world with the 1999 publication of their paper entitled, "The Science of Art: A Neurological Theory of Aesthetic Experience." In it, the scientists claimed to have discovered eight universal principles of artistic experience that "artists either consciously or unconsciously deploy to optimally titillate the visual areas of the brain." Ramachandran used an analogy to explain how he believes science can help us better appreciate the complexities of art: knowing that much of poetry obeys universal laws of rhyme and meter does not make the verses of Shakespeare any less extraordinary. Nor does understanding better how these laws work diminish at all our appreciation of Shakespeare's unmatched mastery of them. In the same way, natural and universal laws govern many aspects of our perceptions. Being a product of these perceptions, art should reasonably be expected to reflect some level of this lawfulness.

The first and most important in Ramachandran and Hirstein's taxonomy of universal aesthetic principles is "peak shift," a well-known effect in animal discrimination learning. When a rat is taught to respond to rectangles and not squares, it will with time respond more strongly to longer rectangles than to shorter ones, having learned the rule ("rectangularity") that the longer the shape—the less squarelike it is—the better. Applying this principle to art, we see a similar amplification of differences in caricatures, where the artist emphasizes the particular, distinguishing characteristics of an individual's features and personality. The result is often considered a striking representation of the person's essence. The scientists believe that exaggeration of this kind is innately pleasing to the observer, because it activates the appropriate neural mechanisms more powerfully than does the person alone.

Ramachandran and Hirstein's second principle is perceptual "grouping" or "binding," where the brain groups similar perceptual effects together, such as grouping together light-colored squares in a painting in contrast to its darker ones. According to the scientists, such groupings stimulate the brain's visual system to send signals to the limbic system (the set of structures in the brain that support emotion, among other functions), producing a pleasant sensation that accounts for the viewer's aesthetic experience of the painting.

Next up is the "isolation" of a single visual component in the artwork, allowing the viewer to give more attention to a particular characteristic (such as form, depth, or color). This idea is equivalent to the famous aphorism "less is more," and helps explain, according to the scientists, why a sketch or outline drawing is more effective as art than a full-color photograph, even though the photo contains more detail.

The concept of "contrast" in art, such as the use of different levels of brightness or color in a painting, also makes the neurologists' list. Their scientific explanation is that the brain finds such contrasts more interesting than homogenous areas because of the useful perceptual information generally located in "regions of change." A painting containing many different contrasts therefore keeps the brain's attention for longer than a more uniform one.

Also helping the brain to stay attentive is the principle of "perceptual problem solving," where the viewer derives pleasure from trying to make sense of an artwork. A picture whose meaning is implied rather than explicit may be more alluring because it forms a kind of "peek-a-boo" that incites the brain's visual system to struggle for a solution, a process that is itself rewarding. Examples of this principle in action include much of Picasso's work and that of the Surrealists, which requires the viewer to actively search for the picture's meaning.

The scientists also point out that most viewers prefer generic viewpoints over unique vantage points. This is because the brain's visual system always picks the most likely interpretation for any visual input it receives, a strategy that works best when there are plenty of vantage points from which the picture can be viewed and deduced. Coincidences in art, such as a tree appearing exactly in between two buildings, are generally not pleasing to the viewer because they are highly unlikely and therefore do not tally with our brain's expectations.

Ramachandran and Hirstein's penultimate principle states that visual puns or metaphors enhance art. This may be because finding connections between things is itself a pleasurable experience. Metaphor in art serves as a form of emotional haiku, communicating complex subjective information in a small number of well-chosen words or images. In visual art, a good painting can often evoke an instant emotional response in the viewer long before it can be rationally dissected and understood. Poets also make use of such metaphors to achieve a similar effect, as in Shakespeare's: "Death, that has sucked the honey of thy breath." The reader's emotional reaction to this line is felt well before he becomes consciously aware of the hidden analogy between the "sting of death" and a bee's sting.

"Symmetry is attractive," conclude the scientists in their survey of art's neurological foundations. Researchers have long recognized the link between symmetry and attractiveness—in one study, researchers found that babies stare longer at pictures of symmetrical individuals than they do pictures of asymmetrical ones. The same relation between symmetry and perceived attractiveness is also found throughout the animal world: for example, female zebra finches prefer males with symmetrically colored leg bands. Ramachandran and Hirstein speculate that symmetry holds our attention because it is a characteristic of biological forms. Our ancestors would have recognized potential predators, prey, or mates in part from their symmetry. Artwork that incorporates symmetrical forms thus plays on this evolutionary "early warning system," to make us particularly attentive to these forms.

Perhaps not surprisingly, Ramachandran and Hirstein's paper has been criticized by various figures within the art world as "reductionistic." Others contend that the aesthetic experience is fundamentally ineffable—a subjective state inaccessible to the rigors of scientific inquiry. The scientists respond by pointing out that their eight principles are far from exhaustive, nor do they wish to eliminate the roles that learning, culture, or experience bring to art's creation. Rather, like Gombrich, they see art and the brain as occupying a symbiotic relationship, in which each helps to enhance the understanding and appreciation of the other.

Of course, no scientific theory can ever really capture the subjective richness and meaning of the artistic experiece. Our perceptions are products of our minds as well as of our brains. Scientists' explanations of how we see the world around us help emphasize the biology of our perceptions, but they also serve to underscore the complexity that gives rise to each person's unique vision. How we see something says at least as much about us as individuals, with our own lived experiences, emotional reflexes, and philosophical views, as it does the neuronal wiring behind our eyes.

8

Food for Thought

Our minds depend on information, just as our bodies require food. Every fact and figure, idea and image, story and statistic helps to shape our memories and our perceptions. In a real sense, then, data is destiny. This is important because, in a time of the internet, blanket advertising, and round-the-clock rolling news, the traffic between our internal and external worlds has never been greater. What are the effects of the modern information age on how we think, learn, perceive, and understand?

The Atoms of Knowledge

To help us understand the nature of our information age, we need first to take a step back and analyze the mechanics of how we acquire, and pass on, our knowledge to and from one another, beginning with an exploration of the very atoms of knowledge: words. The building blocks of books and of every sentence we speak, vocabulary can both feed and frustrate our imagination. The word giveth and the word taketh away. An extreme example of this appears in George Orwell's classic novel *1984,* which describes a society whose words are warped by a propagandistic idiom, Newspeak, a language designed to "diminish the range of thought." The invention of a totalitarian regime, Newspeak makes free thought and free speech impossible because it lacks all potentially troublesome words and ideas, such as "freedom" and "rebellion." The regime's ultimate aim is to strip the language of as many of its words as possible in the belief that something that cannot be said cannot be thought. In Orwell's dystopic vision, all linguistic subtleties are removed, the language reduced to basic dichotomies such as happiness and sadness, pleasure and pain. Purified of all irregularity, in Newspeak "bad" is "ungood," "great" is "plusgood," and "excellent" "doubleplusgood."

Orwell's Newspeak was inspired by a real linguistic project, Basic English, created in 1930 by the British philosopher Charles Ogden. A greatly simplified version of English, it consisted of just 850 words selected to "clarify thought" and simplify the teaching of English overseas, though it found little favor with either government or the general public. President Franklin Roosevelt wrote to Winston Churchill to point out that his famous phrase "blood, toil, tears and sweat" would be translated into Basic English as "blood, work, eye water, and face water." Though initially a supporter of the Basic English project, Orwell emphatically rejected it in his 1946 essay "Politics and the English Language." He reserved most of his fire, though, for modern English itself, which he argued was ugly, stale, and imprecise. His critique still makes interesting reading today. In particular, Orwell singles out "dying metaphors" such as "stand shoulder to shoulder with" and "toe the line," contending that they are often used without understanding what they actually mean. He also attacks "verbal false limbs"—meaningless phrases used to pad out sentences—giving "with respect to" and "make itself felt" among the examples. He derides "pretentious diction," such as the words "expedite" and "ameliorate," used to "dress up simple statements" but which he says results in "an increase in slovenliness and vagueness."

Such examples of ugly and vague language were meant not only to amuse the essay's readers, but also to preface a serious and important point: "if thought corrupts language, language can also corrupt thought." Orwell believed that it was necessary to be constantly on guard against such clichéd expressions, because "every such phrase anesthetizes a portion of one's brain." He closed his argument with a plea to English speakers to employ words thoughtfully and carefully and to be certain of their meaning before using them.

For those wanting advice on how to avoid the linguistic pitfalls he had identified, Orwell gave the following six short rules:

1. Never use a metaphor, simile, or other figure of speech that you are used to seeing in print.
2. Never use a long word where a short one will do.
3. If it is possible to cut a word out, always cut it out.
4. Never use the passive where you can use the active.
5. Never use a foreign phrase, a scientific word, or a jargon word if you can think of an everyday English equivalent.
6. Break any of these rules sooner than say anything outright barbarous.

The "Sokal Affair," as journalists dubbed it, is an amusing example of how language is still often mangled to this day. Alan Sokal, a professor of physics at New York University, had grown exasperated by the frequent misuse of long, scientific-sounding words in many humanities journals to camouflage a total lack of meaningful information. In 1996, he decided to submit a nonsensical, jargon-filled, footnote-heavy scientific paper (entitled, "Transgressing the Boundaries: Towards a Transformative Hermeneutics of Quantum Gravity") to a postmodern cultural studies journal to see whether or not it would be published.

It was. On the day of its publication Sokal announced in another journal that the paper was a hoax: "a pastiche of left-wing cant, fawning references, grandiose quotations, and outright nonsense." A single choice snippet provides a good taste of this: "the [Pi] of Euclid and the G of Newton, formerly thought to be constant and universal, are now perceived in their ineluctable historicity; and the putative observer becomes fatally decentered, disconnected from any epistemic link to a space-time point that can no longer be defined by geometry alone."

Sokal's hoax provoked a firestorm of reaction in both the academic and popular press. Many scientists were sympathetic, but the "postmodern" intellectuals whose own words and phrases Sokal had glued together in his paper were predictably less so. Sokal attributed the acceptance of his parody to the rise in "a particular kind of nonsense" among certain intellectuals, and suggested that his text had not been properly analyzed because its conclusions were politically fashionable.

The use of euphemism—the substitution of vague and obscure words for ones considered blunt or offensive—is another way in which words can be used to conceal as well as communicate meaning (or a lack thereof). Politicians are well known for their careful choice of words: one hundred fifty years before Orwell, the British philosopher Edmund Burke observed that apologists for the French Revolution did not describe things by their common names, calling massacre "agitation," "effervescence," or "excess."

Examples of current political euphemisms are legion. Think "revenue enhancement" for taxation, "economic downturn" for recession, and "unrest" for a riot. Military euphemisms are just as numerous: "collateral damage" for civilian casualties, "defensive strike" for bombing, and "friendly fire" for the accidental killing of one's own or an ally's troops. When President George W. Bush announced plans to send an additional twenty thousand American soldiers to Iraq, he used the word "surge," which several commentators pointed out carried more positive connotations (a dictionary definition of "surge" gives "any sudden, strong increase, as of energy, enthusiasm, etc. . . .") than simply "troop increase."

Such verbal gymnastics are distortions, but they are at least fairly transparent. As the Stanford University linguist Geoffrey Nunberg notes in his book, *Going Nucular: Language, Politics, and Culture in Controversial Times*, much simpler and innocuous-looking words and phrases—such as "choice" and "jobs and growth"—can be the hardest to see through.

Diane Ravitch, a professor of education, explores the relationship of euphemistic language to censorship in her book, *The Language Police: How Pressure Groups Restrict What Students Learn*. Ravitch explains the extraordinarily intricate set of rules used by many educational textbook publishers to censor language and subject material that might be considered controversial or offensive. Surprisingly, she traces much of this censorship to the civil rights and women's movements of the 1960s and 1970s, which campaigned to remove prejudicial language from textbooks, and to open history and literature to neglected voices and points of view. These efforts, Ravitch argues, have been taken to an extreme that favors the blandly inclusive and morally simplistic over complex analysis and the free play of ideas.

Many of the examples in Ravitch's book seem to border on farce: a legend about dolphins was considered problematic by one publisher because it was seen as reflecting a regional bias against children who do not live near the sea; a passage about owls was rejected from a standardized test because the birds are taboo for Navajos; Aesop's fable, "The Fox and the Crow," was marked out as sexist because a male fox flatters a female crow (the sex of the animals had to be changed before the story could be accepted). Ravitch's research indicates that such examples are far from the exception; most tests and textbooks used in American schools are in fact governed by similarly labyrinthine and frequently absurd "sensitivity-and-bias" guidelines. Pressured by interest groups at both ends of the political spectrum, textbook writers attempt to please everyone by bowdlerizing potentially offensive passages, while cramming the pages with innocuous material.

The result, as Ravitch notes, is dry, boring, and insipid content lacking any overarching narrative that might inspire in students a love of history or literature. All diversity is expunged, reducing personalities to interchangeable beings whose differences are ignored—hardly a healthy basis for encouraging perceptive, critical thought. Preventing children from being exposed to a wide range of words and ideas (including "white collar," "unmarried," "widow," "addict," "landlord," "brotherhood," "yacht," "cult," and "primitive") limits their access to the world's complexities and stifles their imaginations. The irony in all of this is that the censors' books are themselves guilty of stereotyping, by mandating that students only be exposed to material that fits their presumed experience.

Some language experts believe that the words we use affect a reader's or listener's thought and perception far less than the preexisting conceptual framework through which each person makes sense of them. Debate, according to this view, is not about finding the right words but the right "frames." This argument—known as "framing"—has been especially advanced and popularized in recent years by George Lakoff, professor of cognitive linguistics at the University of California, Berkeley. For example, in the term "tax relief," Lakoff argues, the word "relief" evokes the idea of someone afflicted with something harmful or painful that should be taken away. The relief "frame" involves a hero (the reliever of the pain), a victim (the afflicted person), a crime (the affliction), a villain (the cause of the affliction), and a rescue (the relieving of the pain). In these terms, tax is seen as an inherently negative thing and its reduction as a heroic act for which the taxpayer should be grateful. This particular perspective on taxation is reinforced each time that it is repeated by politicians and in the media, Lakoff contends.

For liberals and others who do not agree with this view of taxation, Lakoff suggests framing the issue of tax differently, by using an alternative vocabulary that stresses alternative values, such as fairness, community, and cooperation. Accordingly, opponents of "tax relief" should talk about taxation as "the price of civilization" (as did Oliver Wendell Holmes) or as your "dues" to democracy, opportunity, and access to the infrastructure that makes the acquisition of wealth possible.

Lakoff's advice is firmly pessimistic, suggesting that facts will always be trumped by a person's worldview. The linguist Steven Pinker has criticized Lakoff's arguments as "cognitive relativism" that make math, science, and philosophy into "beauty contests" rather than attempts to understand the nature of reality. In contrast, Michael Silverstein, professor of anthropology, linguistics, and psychology at the University of Chicago, uses a different model for persuasion in his book, *Talking Politics*. Silverstein argues that language is only truly effective when it marries persuasive style with real substance. He cites Abraham Lincoln's brilliantly moving 272-word Gettysburg Address as evidence for the power of words when imbued with genuine understanding and conviction. As Silverstein reminds us, language at its best becomes an instrument of reason and reflection rather than pandering or prejudice.

Grooming with Words

Of course, not all words are pressed into the service of persuasion, or the communication of deep thoughts or high ideals. I will freely admit bafflement at the phenomenon of gossip—my own mind just does not work that way. I need not feel foolish, however: scientists have been just as puzzled over the question of what gossip is actually for. What role does it serve, for the person and for society at large?

Gossip has in recent years become the subject of much academic debate among a range of fields including social psychology, anthropology, evolutionary psychology, and sociolinguistics. Though the word itself carries negative connotations—perhaps not surprising when we consider the grief negative and unsubstantiated rumors can cause—research has uncovered many positive social and psychological reasons for gossiping: from group bonding to the reinforcing of shared values.

There is also evidence that gossip is a deeply rooted human instinct. Robin Dunbar, professor of psychology at the University of Liverpool, points out in his book, *Grooming, Gossip, and the Evolution of Language*, that most people devote around two-thirds of their conversations to gossip—"the natural rhythms of social life." Dunbar believes gossip is the human equivalent of primate grooming, where the animals spend hours grooming each other's fur as a form of social bonding. Primates, who live in groups, do this as a way of helping to keep their community together. Dunbar suggests that humans evolved to use language instead because it is less time-intensive and allows the individual to do more than one thing at a time.

Numerous studies by researchers seem to back up the idea of gossip as a primarily positive and socializing activity: one study showed that only 5 percent of time spent gossiping involved criticism and negative evaluations, the bulk of conversations being focused on "who is doing what with whom" and personal social experiences. Another study showed that just ten minutes of social conversation per day was as effective as doing a crossword for boosting memory and mental performance.

A less healthy form of gossip is the urban myth or urban legend—a story of doubtful authenticity that is nonetheless presented as being true (often the teller claims the related story happened to a friend of a friend or distant relative). The folklorist Jan Harold Brunvand, author of *The Vanishing Hitchhiker: American Urban Legends and Their Meanings*, argues that urban legends are a modern form of narrative folklore, of storytelling that has gone on between people since time immemorial. Their enduring popularity in an age of widespread literacy and media rests in their ability to reflect many of the hopes, fears, and anxieties of our time. No one knows where most of these legends originate or who invented them—most leads pointing to possible authors or original events prove false upon further study. The most persistent urban myths succeed nonetheless in capturing the public's imagination because they have a solid narrative structure, vague plausibility, and an underlying message or moral. One such long-running legend has it that alligators live in the sewers of New York, flushed by apartment dwellers who had brought them from Florida vacations as babies. The facts suggest this is pure bunkum: New York is too cold for alligators, who live at temperatures between 78 and 90 degrees, not to mention the disease-transmitting bacteria found in sewers.

A variation on these cultural legends is the "suburban myth"—common misconceptions that are repeated by experts and nonexperts alike. One such myth, particularly relevant to the subject of this book, is that people only use 10 percent of their brains. There is no good evidence to support this idea and plenty against it. For example, several neurological disorders, such as Parkinson's disease, cause devastating disabilities yet damage much less than 90 percent of the brain. It is also highly unlikely from an evolutionary viewpoint that larger brains would have developed if there were no advantages related to greater brain size.

A further example of such myths, and the damage they can cause, is the claim that MMR (measles, mumps, rubella) vaccinations can cause autism. There is no sensible reason to believe that any vaccine can cause autism. Because the MMR vaccine is first given to children at twelve to fifteen months, around the age when autistic symptoms are often first noticed by parents, any perceived correlation is due to coincidence rather than cause and effect. In fact, millions of children have been vaccinated over many decades without ill effects. The British medical journal, *The Lancet*, published the only "evidence" purportedly linking the MMR vaccine and autism in 1998. The article, by gastroenterologist Andrew Wakefield, appeared with an editorial discussing concerns about the study's validity. Wakefield and colleagues speculated that the MMR vaccine might cause bowel problems that lead to a decreased absorption of essential vitamins and minerals, resulting in developmental disorders such as autism. No scientific analyses were reported to support the theory, however, and there remains to this day no evidence that autism can be caused in the way Wakefield claims. A recent study of 498 cases of autism spectrum disorder cases in London found no jump in diagnoses following the introduction of the MMR vaccine in 1988.

The vast majority of doctors reject Wakefield's conjectures. Michael Fitzgerald, a British general physician and father of an autistic child, spoke for many when he charged that Wakefield had "opted out of medical science to join the world of pseudo-scientific dogma, media celebrity and populist campaigning." Unfortunately, none of this prevented Wakefield's claims from being widely and often uncritically reported in the media, leading many parents to decide not to have their children immunized against the triad of potentially fatal childhood diseases. As a result the occurrence of measles—a highly infectious and dangerous condition—in the United Kingdom has risen significantly in recent years. Nearly a thousand cases were reported in 2007, the highest total on record. The year before, a thirteen-year-old boy became the first person to die from measles in Britain in a generation; a stark reminder of how dangerous the circulation of inaccurate and misleading information can sometimes be.

Remarkably, researchers have discovered a phenomenon within the brain that would account for many people's willingness to believe the unsubstantiated or illogical statements they read or hear. Neuroscientists Sam Harris, Sameer A. Sheth, and Mark S. Cohen used fMRI (functional magnetic resonance imaging) to scan the brains of fourteen adults at the University of California's Brain Mapping Center. The scientists presented the volunteers with a series of statements that were written to be plainly true, false, or undecidable. The volunteers pressed a button to indicate their belief, lack of belief, or uncertainty about each of the statements, which were mathematical ("the number 62 can be evenly divided by 9"), factual ("the Dow Jones Industrial Average rose 1.2 percent last Tuesday"), and ethical ("It is bad to take pleasure at another's suffering").

The results were striking. Evaluating the different kinds of statements involved different regions of the brain, indicating that belief, disbelief, and uncertainty generate separate neural pathways. The research also showed that the volunteers responded more quickly to statements they believed true than for those they did not believe or were uncertain about. This finding supports the idea put forward by the seventeenth-century Dutch philosopher Benedict Spinoza, that simply understanding a statement entails a tacit acceptance that it is true, whereas disbelief requires a subsequent process of rejection. The researchers suggest that understanding a statement or idea may be equivalent to perceiving an object in physical space, as people generally accept appearances as reality until proven otherwise.

Living Literature

In order to better evaluate the latest advances of the modern information age, it is enlightening to look back to the ancient one. For millennia, humans have spent much of their lives acquiring and transmitting knowledge from and to one another. Long before the written word, many cultures recited epic songs, poems, myths, and proverbs from memory to pass what they knew from generation to generation. These societies were more concerned about the very survival of information than about its mangling or misuse. If a piece of learning was forgotten, it was lost forever.

Each culture's "oral literature" was often learned and recited by a designated speaker or singer, who learned his craft from other speakers. Public recitations were common, during which the speaker told tales, sang songs, and recounted dramatic scenes. These living memories helped to educate individuals and provide a rich personal sense of identity, community, and continuity. The songs and stories were highly structured, rhythmic, and repetitive—characteristics that aided both the teller's and listener's memories. The tales also revolved around vivid characters and events that pulled the listener in and held his attention. The speaker commonly improvised without distorting the main points of the story, allowing him to vary his accounts according to his audience's taste and level of interest, and to continually hone them using their feedback and new learning.

Australian aborigines provide a good example of the subtlety and complexity of oral traditions. Central to theirs is a high respect for communal memory, born from their belief that the past exists in an eternal present. The Aborigines' narratives (known as "song lines") describe their relationship to the land, helping to reinforce their knowledge of local geography, such as food routes and the location of water holes. Different clans will meet to sing these lines to one another, in order to exchange detailed information about the territories in surrounding regions.

Elias Lönnrot, a nineteenth-century doctor and folklore scholar, probably did more than any one man to rescue the wealth of oral literature in Europe. In a dozen expeditions, Lönnrot traveled across Estonia, Lapland, and Russian Karelia in search of "runo"—the ancient sung poetry native to these regions—organizing the thousands of lines of various songs that he found into a single body of poetry. Lönnrot published his magnum opus, the *Kalevala*, on February 28, 1835, a date celebrated in Finland today as Kalevala Day—the birthday of Finnish culture. The myths that Lönnrot saved from oblivion spoke of a past where unseen gods and spirits governed men's lives. There were creation stories, legends dealing with light and darkness, fertility and death, and detailed descriptions of the land's animals, plants, and seasons. Lönnrot's work eventually inspired the Finnish people to establish their own nationhood and elevate Finnish to a national language. To give a small idea of its rhyme and power, here are the Kalevala's opening lines in English translation:

> I am driven by my longing,
> And my understanding urges
> That I should commence my singing,
> And begin my recitation.
> I will sing the people's legends,
> And the ballads of the nation.
> To my mouth the words are flowing,
> And the words are gently falling,
> Quickly as my tongue can shape them,
> And between my teeth emerging . . .

The importance of oral literatures declined following the invention of writing in ancient Sumer around 3200 BC. With the advance of the written word came scribes, bureaucrats, and the origins of modern civilization. The advance allowed news and ideas to be carried to distant places as well as the creation of records, liturgies, and other documents. Though many societies saw writing as a divine gift, they were as aware of its costs as of its benefits.

Preliterate speakers populated their minds, and those of their listeners, with complex mental tapestries of images and ideas. Their voices used accent, pitch, tone, and emotion, as well as words, to tell their tales. As the role of writing eclipsed that of remembering, many regretted that it had made memory a forgotten art. No less an intellectual figure than Plato shared in this ambivalence towards writing, using the word "pharmakon" for it, meaning both "medicine" and "poison."

In spite of the new opportunities that writing created, literacy was slow to gain traction in many societies. Those who could read and write, mostly monks and the nobility, seemed more than happy to keep their knowledge to themselves. Great literary works were written and circulated among these classes but were out of the reach of the common man.

All this changed with the invention of the printing press by Johannes Gutenberg around 1440, inspired by the wine presses of the Rhine Valley. Gutenberg's method of printing created a literacy revolution, as books became far more numerous and affordable to the masses. By 1500, there were more than one thousand printing shops across Europe. The new accessibility to the written word helped foster unprecedented social and intellectual debate and the rapid development in the sciences, arts, and religion. Increasingly, books became a part of people's lives.

The explosion in books gradually increased the role of libraries as repositories of written knowledge. Aristocrats of the period were urged to open their private collections to the public. The librarian Gabriel Naudé argued that collecting books served little purpose if they were not available to those who could make use of them. He issued guidelines to new libraries, stating that their books should be lent for limited periods of time to "persons of merit and knowledge" with records kept for each borrowing.

As the quantity of readily available information grew, so too did the encyclopedic impulse to catalog it. In 1704, John Harris compiled what is widely regarded as the father of the modern encyclopedia, the *Lexicon Technicum*. Written in clear text, with bibliographies and cross-references, it became the standard for all that succeeded it. The first of these, Diderot's *Encyclopédie* (issued between 1751 and 1772), was so extensive that it prompted the writer Voltaire to say of it: "this vast and immortal work seems to reproach mankind's brief life span."

Most famous of all, however, is the *Encyclopædia Britannica*, the oldest English-language encyclopedia still in print, whose first edition was produced between 1768 and 1771. Rather than rely solely on its own team of writers, Britannica actively sought out the contributions of a wide range of important contemporary figures, including Sir Walter Scott, Sigmund Freud, Albert Einstein, Marie Curie, Leon Trotsky, Harry Houdini, G. K. Chesterton, and H. L. Mencken.

Informal Information

Today's encyclopedia of choice can rightly claim many more contributors than any other in history. Wikipedia, the online encyclopedia that can be edited by anyone, has grown exponentially since its inception in 2001. In a few clicks of a mouse, readers can learn about such diverse topics as "heavy metal umlauts," "toilets in Japan," "Saddam Hussein's novels," and "chess-related deaths." Wikipedia's English-language version alone boasts more than two million articles, consisting of over a billion words. With such a wealth of freely available information, it is little wonder that it now ranks among the world's ten most visited websites.

Uniquely among encyclopedias, Wikipedia does not require an article writer or editor to have any credentialed knowledge in a particular field. Its open structure allows anyone to contribute to any of its articles. Wikipedians claim that this radical openness works because a diverse group of people is usually smarter than any one of its members. In this way, they hope to create an encyclopedia as good, or even better, than one written by experts.

But there are significant problems with the Wikipedia model. For one, it is far less democratic or meritocratic than often believed. For example, in 2007, researchers at the University of Minnesota released the results of a study showing that 10 percent of editors contributed 86 percent of the site's edits, and just 0.1 percent contributed 44 percent. This small clique of Wikipedia's super editors control the vast bulk of the site's content. Anonymity is another problem. Many editors contribute under aliases, which a small number have used to carry out their own agendas and evade responsibility for what they create. One set of hoaxers created an article on Henryk Batuta, who the article stated was born in Odessa in 1898, participated in the Russian Civil War, and was an ally of Ernest Hemingway during the Civil War in Spain. The article was available for fifteen months, and cross-referenced by editors to seventeen other articles before it was finally revealed that Henryk Batuta never existed. The hoaxers claimed their intention was to warn Web users against "swallowing information" without checking it.

What happens when a knowledgeable professional spots an error in one of Wikipedia's articles and attempts to correct it? William Connolley, a climate modeler at the British Antarctic Survey and global warming expert, tried to correct inaccuracies in an article on global warming, only to be accused by a skeptic of having "singular" and "narrow" views on climate change. His credentials given no more weight than the postings of his anonymous rival, Connolley was eventually limited to making a single edit per day as a result of the fallout.

A defense often given for the Wikipedia model is that it relies on the "wisdom of the crowds"—a term popularized by journalist James Surowiecki's 2004 book of the same name. Surowiecki believes that "the many are often smarter than the few," citing the example of the popular TV show, *Who Wants to Be a Millionaire?* When called upon, experts answer a contestant's question successfully two-thirds of the time, yet the studio audience members manage the right response over 90 percent of the time.

How can this be? Imagine a contestant is given the following question:

"Which of the following is not a capital city?"

A: Prague
B: Quito
C: Sydney
D: Edinburgh

The correct answer is C; the capital of Australia is Canberra. Now let's say that the studio audience's collective wisdom breaks down as follows:

14 know the correct answer.
20 can eliminate two of the options.
30 can eliminate one of the options.
36 do not have a clue.

Here then is how this audience will vote:

The 14 who know the right answer will vote C.
10 of the 20 who are left with two possible choices will vote C.
10 of the 30 who are left with three possible choices will vote C.
9 of the 36 without a clue will vote C.

C easily comes first with 43 votes. Even though only one in seven knew the answer for sure, nearly half the audience came up with the right result. By aggregating their individual pieces of knowledge they were able to help the correct answer rise to the surface.

It is far from clear, though, whether Wikipedia could operate in a similar way. After all, compiling meaningful bodies of knowledge is very different from answering multiple-choice questions. Surowiecki points out that a crowd's wisdom depends on a mechanism for assembling the individual, independent decisions together, but this is not possible for an encyclopedia entry. Imagine, for example, that we ask one hundred people with varying amounts of knowledge about Islam to write down what they know. It is unclear how these diverse contributions could be tabulated, like the educated guesses for a quiz question, to arrive at a coherent and comprehensive summary. In reality, Wikipedians rely on far less sophisticated techniques—compromise and consensus—for the evolving composition of their articles. But is this method as reliable as the work of traditional encyclopedias? In 2005, journalists for *Nature,* the prestigious scientific journal, carried out one widely publicized study in an attempt to answer this question. The researchers sent forty-two pairs of unattributed articles on science topics, half taken from Wikipedia and half from the *Encyclopaedia Britannica*, to a range of experts asking them to rate the articles' accuracy. From the feedback received, the researchers concluded that Wikipedia was comparable in accuracy to its more traditional rival.

But there were serious flaws in *Nature*'s study. Its reviewers claimed to have found omissions in a number of *Britannica*'s entries, yet in several cases only segments of the actual articles had been reviewed; one reviewer was sent only the 350-word introduction to a 6,000-word article on lipids. *Nature*'s journalists also failed to distinguish between minor inaccuracies and major errors, giving equal weight to every mistake found. In fact, the study's reviewers reported that many Wikipedia articles were "poorly structured and confusing," making comparisons between articles in this way virtually meaningless.

The idea that a large, coordinated group can outperform a talented expert is also questionable. One counterexample is the 1999 online match between world chess champion Garry Kasparov and thousands of amateurs, in what was billed as "Kasparov versus the World." A public vote, to which anyone could contribute, decided each of the "world" team's moves. In all, some fifty thousand people from more than seventy-five countries took part. Yet after four months and sixty-two moves, it was Kasparov—not the crowd—who claimed victory.

Wikipedia's success seems to owe much to a collective loss of faith in experts and with it a more casual relationship to truth. In deference to this cultural shift, many other websites have blurred the distinction between those with relevant knowledge to share and others who learn from them. Take, for example, the web pages for the *Guardian* —one of the biggest-selling newspapers in Britain. Its articles are researched and written by experienced journalists or leading public figures and published online. Readers are then invited to add their own views to the end of the story. In today's *Guardian* there is an article on the issue of the relationship between China and Tibet, written by Václav Havel, former president of Czechoslovakia, accomplished writer and dramatist, and a recipient of the Presidential Medal of Freedom and the Ambassador of Conscience Award. It is far from clear what value is added to Havel's words by anonymous posters tacking their own comments to the end of the former president's.

Television news programs, too, are changing the definition of what does and does not constitute meaningful information. A 2004 study by researchers at the University of Cardiff indicates that rolling news broadcasts in particular do a poor job of providing informative, relevant content in the stories they report. One reason is their unquenchable thirst for comment from politicians and other public figures, who in turn hire public relations agencies to help them deal with the unrelenting demand of twenty-four-hour news—leading, ironically, to less clarity, not more. Conscious of the need to keep viewers from switching channels, rolling news programs also feature interviews that, the researchers argue, are designed more to elicit a provocative statement—that can be subsequently discussed and dissected at length by commentators—than to improve the public's understanding of an issue or event. Also criticized is the amount of time wasted in "breaking news" stories that are often in fact nonevents: discussing the imminent return of the British football team from an international competition while showing "live scene" footage of an empty airport runway, for example. Sometimes such trivia is in decidedly bad taste, such as the regularly updated TV reports of a dying Pope John Paul II in 2005, which described in gory detail his progressive heart and kidney failure.

The very nature of the twenty-four-hour news format, dedicated to rapid roundups of running stories, means there is actually less time, not more, for explaining or investigating the news in the way that a single flagship evening news program can. As a result, the researchers conclude, those who watch rolling news are actually likely to be less well-informed than those who watch a single nightly program that provides some analysis and context.

Ad Attack

The modern proliferation in information extends outside the home as well. Advertisements are omnipresent, found in even the most unlikely of places, inlcuding gas pumps, elevators, turnstiles, mailboxes, trash cans, and even stuck to fresh fruit. Victims of their own success, advertisers are forced to find ever more inventive (and intrusive) ways to promote their products, as public spaces suffer increasingly from "ad clutter." Three recent examples: the California Milk Processing Board put up a bus shelter in San Francisco that smelled like cookies (until ordered to take it down by city officials). McDonald's logo has appeared on report cards disseminated to elementary school children in Florida. Even outer space is not safe: in 2000, the Russian space program launched a rocket displaying a thirty-foot ad for Pizza Hut.

Critics warn of the growing bombardment of the senses with commercials, describing it as "ad creep," and argue that it uglifies public places and encroaches on individuals' privacy and right to personal space. Even some people in the advertising business are worried that too much advertising is actually counterproductive, with consumers simply "switching off" from such messages due to ad overload.

Commercials thrive on bold imagery and vague language. For example, a detergent that claims to "leave dishes virtually spotless" is an example of a "weasel word"—a modifying word ("virtually") that negates the claim that follows. Other common weasel words in advertising include "can be," "up to," "as much as," and "looks like." A further example of vague language in commercials is the unfinished claim, where the ad claims the product is better without stating the nature of the comparison: "X gives you more" begs the question "More what?"

Trickier to spot can be the dubious claims of efficacy made in many ads, especially when accompanied by impressive-sounding statistics. Frequently these are intended to demonstrate that a product's users judge it as highly effective, as in: "86 percent of women said they saw a difference in just seven days." The number of people asked, however, is usually very low—less than a hundred—and it is never clear how the data is collected.

Peer pressure is an especially insidious form of advertising, known technically as the "ad populum" fallacy: everyone else is doing it/eating it/wearing it/driving it, so I should, too. Of course, this is not in itself a reason for you to do so as well, but advertising slogans such as, "the country's number-one selling . . ." tap into this impulse. Teenagers are particularly vulnerable to peer pressure, since they are still in the process of acquiring their values and developing a self-image. A 2006 study from the University of Connecticut confirmed that teenagers and young adults who see more alcohol advertising are indeed likely to drink more. The researchers interviewed four thousand Americans aged between fifteen and twenty-six about their drinking habits and viewing of ads. They discovered that each additional alcohol ad seen per month was associated with a 1 percent rise in the average number of consumed drinks. Better news comes from more recent research that shows that adolescents who learn how to critically evaluate ads for alcohol are more likely to resist using it.

Advertisers use other "tricks of the trade" besides words and images to influence consumers' perceptions and choices. Psychologists at Northwestern University found that simply adjusting how a product is presented—what they call the "perceptual focus effect"—can dramatically alter people's preferences for it. In tests, subjects were asked to choose between two sofas— one that had softer cushions compared to another that was more durable. The more durable sofa won out—only 42 percent chose the one with soft cushions. However, when the two sofas were grouped with three other sofas that were also more durable, the preference for the sofa with soft cushions jumped to 77 percent. Manipulating the context in which we view a product in this way serves to visually emphasize its distinctiveness, making it more appealing to potential buyers.

Stanford neuroscientist Brian Knutson believes that these results occur because people's decision making is more intuitive and emotional than rational, a theory he tested in the laboratory. Knutson gave subjects twenty dollars and, while they were inside a scanner, presented them with a series of pictures of products with prices displayed. The subjects were then given the option to buy each item shown. As subjects looked at products they preferred, the scanner indicated activity in the nucleus accumbens—a brain region involved in anticipating pleasant outcomes. When the subjects thought the price of an item too high, however, the scanner showed increased activity in the insula cortex—an area involved in anticipating pain. By looking directly inside people's heads before they make certain decisions, Knutson contends, it is possible to see what emotions are being provoked by a product and predict what they will do (to buy or not, in this case).

Brain imaging also provides an explanation for the power of marketing brands. In our daily lives, our brains save time by paring down the decision-making process, setting up shortcuts that make its analysis more efficient. These shortcuts mean that factors like previous experience are far more important to a consumer's decision than careful value judgments or risk-versus-benefits calculations. Brands are familiar, and for this reason, they sell.

Still, purchasing behavior, like any other, is complex—a mixture of the personal as well as the neurological. And as more of this kind of research is published, the hope is that its details will filter into the public consciousness. Knowing as much as advertisers do about how our brains work will be an important part of making good choices as customers, and avoiding bad ones.

Overcoming Information Overload

Our world is generating more information with more resources and technology now than at any time in history: through TV and radio programs, cell phones, magazines, email, websites, blogs, and other media. There is no doubting the benefits that the free and plentiful flow of information has brought to our lives, but as many people are finding out: there really can be too much of a good thing.

Being overwhelmed by a continuous maelstrom of information can be just as damaging to our minds as having too little of it; both extremes dampen down careful, reflective thinking, the ability to make meaningful connections between disparate facts or ideas, to gain genuine understanding of complex issues and events, and to make sense of ourselves and the world around us. The modern "toomuchness" of information is eroding both the vigor and rigor of our mental lives.

In numerous studies, psychologists give support to the idea that too much information can be harmful to our brains. In 1997, journalist David Shenk touched on many of these concerns in his book, *Data Smog: Surviving the Information Glut*, arguing that modern forms of information were multiplying faster than our ability to process them, leading to "infoglut" and detracting from our quality of life. British psychologist David Lewis describes the negative effects of data smog—from insomnia to poor concentration—as "information fatigue syndrome," and business executives in his case studies show symptoms ranging from irritability to heart problems and hypertension. Dr. Lewis's studies also show that workers struggling with an excess of information are more likely to make mistakes or misunderstand coworkers and orders, and to work longer hours in an attempt to keep up with the flow of new information.

When faced with a plethora of information, many people try to multitask, but scientific research suggests that this does not help. René Marois, a neuroscientist and director of the Human Information Processing Laboratory at Vanderbilt University, measured how much efficiency is lost when two tasks are carried out at the same time. The first task involved pressing the correct button in response to one of eight sounds, while the second asked subjects to say the correct vowel after seeing one of eight images. When given the tasks one at a time, the participants' performance for each task was not significantly different. However, when asked to perform the two tasks simultaneously, the subjects significantly slowed in their performance of the second one.

One of the most common negative side effects of information overload is distraction, which costs people and companies time and efficiency. Eric Horvitz, a research scientist, and his coresearcher Shamsi Iqbal carried out a study to evaluate the effect that distractions like email or web surfing have on a worker's ability to perform serious mental tasks, such as writing reports or computer code. They found that responding to an email or instant message slowed workers down considerably: on average each needed around fifteen minutes after the interruption before settling back into productive work. The initial distraction often snowballed as the workers replied to other messages or browsed websites. One estimate for the financial cost to the American economy of such lost productivity puts the figure at as much as $650 billion per year.

The hope of many is that technology can help produce a solution to the problem it helped to create. Anti-spam filtering is a good example of this. Spam—the name for the indiscriminate sending of unsolicited bulk messages—is a particularly frustrating problem for web users, with estimates suggesting it accounts for four-fifths of all emails. The content of these messages is often offensive, or contains scams to trick the greedy or gullible. Fortunately, email filters and quarantine folders can significantly reduce the amount of spam you might otherwise receive.

To help you avoid cyber junk mail, internet advisers recommend that you never reply to any spam message, even to ask to be removed from the sender's contact list. You should also avoid using your actual email address when posting a message on a newsgroup. Think of your email address as you would your home address and be just as wary of giving it to others.

Washington state computer scientist Gordon Bell has devised a more extreme technical solution to the problem of information overload. For the past decade, Bell has been creating a vast digital archive of his life on a computer he calls his "surrogate brain." A tiny camera around his neck captures minute-by-minute images of his daily experiences, while an audio recorder tapes the contents of his every conversation. His archive includes more than 100,000 emails, 58,000 photos, thousands of recorded phone calls, and logs of every website he has visited since 2003. This "lifelogging" experiment has won both admirers and detractors. Some view him as a pioneer for a not-too-distant future of virtual memories that will make light work of our data deluge. Frank Nack, a computer scientist like Bell, disagrees, emphasizing instead the importance of forgetting. Forgiving someone, he points out, requires the ability to forget particular elements of our past. Others worry that recording our lives would make those around us cagier and less natural, feeling as though they were always performing for the camera.

Another problem with "surrogate brains" is the negative effect they have on our real ones. In 2007, neuroscientist Ian Robertson interviewed three thousand adults, asking them for standard personal information. He found that less than 40 percent of those under the age of thirty could remember a single relative's birthdate. Even more surprising, fully a third had to rely on their mobile phones to tell them their own telephone number.

Far more significant then forgetting such details is the impoverishment of our self-understanding that comes from comparing the brain to a computer's data storage system. As we saw in chapter 3, our memories are not bits of data but complex patterns of story, imagery, and emotion. The poet Derek Walcott makes a similar point in his 1992 Nobel Prize lecture, where he compares our memories to fragments of a cherished vase that we lovingly piece back together. It is the very act of putting the pieces back together, Walcott suggests, that helps us to love.

Technical solutions can only ever be a small part of the solution to the problems of information overload. Personal decisions and actions that seek to take control of how we acquire information and knowledge are much more important. Establishing boundaries and deadlines is perhaps the simplest way of doing this: turning off your work cell phone outside of office hours, for example, and deciding to check email no more than once per hour.

Learning how to search systematically for required information is a valuable way to avoid wasting lots of time and energy. Putting single words into search engines is never as efficient as using multiple specific terms and punctuation to guide your search. For example, typing "first novel" and "Sherlock Holmes" (the quotation marks tell the search engine to look for the words within them as a complete phrase) produces 70,000 results and the answer—*A Study in Scarlet*—in the very first one, compared with 320,000 results without the quotation marks.

Step away occasionally from your computer and into your local library, where information is stored in a clear, sophisticated layout that allows rapid access to thousands of books on hundreds of different subjects. Navigating your way around a library's shelves is a useful, if sadly undervalued, skill. Most libraries use a system called the Dewey decimal classification scheme (named after the librarian Melvil Dewey) to organize their nonfiction books into specific categories. According to the Dewey system, all books on the same subject are found in the same area, while books on related or similar subjects are found nearby. In this way a book's similarity or content's relation to another is represented in a spatial architecture—the more closely that two books are related, the closer together they are on the library's shelves. The system gives each book a code, allotting it to one of ten major categories according to an intricate yet beautifully intuitive classification:

Section 000–099: General: encyclopedias, directories, books of facts and records, IT, and the paranormal.

Section 100–199: Philosophy and Psychology: books about ideas and the mind.

Section 200–299: Religion: religious systems and beliefs.

Section 300–399: Social Issues: how society works and functions.

Section 400–499: Languages.

Section 500–599: Science: math, astronomy, physics, chemistry, nature, plants, birds and animals, the weather.

Section 600–699: Technology: machines and inventions, electronics, medicine and the human body, farming, pets, food and cookery.

Section 700–799: The Arts: drawing, painting, photography, music, dance, theater, hobbies, and sports.

Section 800–899: Literature: poems, plays, and critical works.

Section 900–999: History, geography, and biography.

These sections are then divided into subsections, so for example:

Section 700–709: The Arts: general books.

Section 710–719: Town Planning.

Section 720–729: Architecture.

Section 730–739: Sculpture.

Section 740–749: Drawing/Decorative Arts.

Section 750–759: Painting and Paintings.

Each of these subsections is in turn subdivided into specific topics:

730: Sculpture.
731: Processes, forms, and subjects of sculpture.
732: Sculptures up to 500 AD.
733: Greek, Etruscan and Roman sculpture.
734: Sculpture from c. 500 to 1399 AD.
735: Sculpture from 1400.

The subjects can become even more specialized by affixing decimal points to the numbers (the more numbers after the decimal, the more specialized the topic):

739: Art Metalwork/Jewelry.
739.2: Work in precious metals.
739.27: Jewelry.
739.274: Jewelry making techniques.

Memorizing every number in the Dewey system is not necessary: libraries use alphabetical subject indexes, allowing the searcher to look up a topic and find the number next to its name. Noteworthy, too, is how the books' arrangement naturally unfolds, beginning with the general (encyclopedias, dictionaries), then moving to systems of thought (philosophical, religious, and social) before tackling the sciences and humanities. The subsections are likewise organized intuitively, from the general to the more specific, occasionally using devices such as chronology to help give the information a meaningful arrangement.

Dewey's system is a marvel of organization, but I have given detailed examples here in order to make an important philosophical as well as practical point. Information is meaningless unless it can be made sense of, and to do that it requires an internal system of thought and ideas that can provide context and relate it to other information we have already learned.

Many people lack a coherent worldview with which they can evaluate and assimilate new information. The problem of information overload, therefore, may not be the quantity of it but our inability to know what to do with it. One possible explanation for this is the common confusion between information and ideas. In his book, *The Cult of Information*, history professor Theodore Roszak makes the point that the mind thinks with ideas, not information. Ideas are of primary importance because they define, make sense of, and create information. Roszak goes further still by arguing that the greatest ideas, such as the Founding Fathers' "all men are created equal," do not contain any information at all. Rather, such ideas are the result of an innate human sensibility that reaches beyond strings of data to recognize and synthesize transcendent patterns of thought. A personal worldview then helps put information back into perspective, giving it an intuitive place in our minds like the books in a library.

Creating such a system of ideas for your mind starts with the cultivation of a healthy curiosity about yourself and the lives and the world around you. Never stop asking questions, even if the answers seem far removed from your ability to immediately glimpse or grasp them. Find joy in learning. Exercise your innate desire to discover truths about our existence, something I believe everyone possesses. Understand, too, the enormous difference between knowing the name for something and really knowing it. Physicist Richard Feynman often quoted this argument made by his father:

> "See that bird?" he says. "It's a Spencer's warbler." (I knew he did not know the real name.) "Well, in Italian, it's a Chutto Lapittida. In Portuguese, it's a Bom da Peida. In Chinese it's a Chung-long-tah, and in Japanese it's a Katano Takeda. You can know the name of that bird in all the languages of the world, but when you're finished, you'll know absolutely nothing whatever about the bird.

You'll only know about humans in different places, and what they call the bird. So let's look at the bird and see what it's doing—that's what counts!"

Use your imagination as much as possible, especially in "thought experiments" that force you to think about the consequences of something being true. Take, for example, the urban myth of the alligators in New York's sewers that I described near the start of this chapter. Consider for a moment the consequences of this actually being true. As one official noted wryly, if those alligators really did exist, sewage workers' unions would be demanding a pay increase to compensate for the extra risk involved in their work.

Perhaps most important, treat each new piece of information you read or watch or hear as a potential piece in a puzzle, rather than as simply an end in itself. Acquiring information is not the same as learning, or thinking, or living for that matter. Bits of information are what we use to build reflections, evaluations, and understanding in our minds. Like each one of us, these dots of data make most sense when they contribute to something greater than themselves.

9

Thinking by Numbers

The beauty of mathematical thinking is that you can do it anywhere. All you need is a little peace and a lot of patience. A willingness to look beyond what conventional wisdom says helps, too. For an example of what I mean, let's begin with a familiar question: what is the maximum number of times that a piece of paper can be folded in half? Until a few years ago the answer found in many books was seven or eight. Some teachers even folded sheets of paper in their classes to help demonstrate this. But, in 2001, a high school student from Pomona, California, proved them all wrong.

Unconvinced by what her math teacher had told her, Britney Gallivan decided to put the famous paper-folding limit to the test. Gallivan wondered whether the number of times a piece of paper could be folded might be related to its length and thickness. She noted that while a letter-size piece of paper can be folded into six successively shrinking halves, larger sheets could go to seven or even eight folds. It appeared then that the ratio of a sheet's thickness to its length is what, in fact, determined how many times it could be folded.

For example, a particularly thin letter-size sheet has a length around 10,000 times greater than its thickness. After 1 fold it is 2,500 times longer than thick, after four folds it is 39 times longer, after 6 folds just 2.5 times. A seventh fold is impossible because the paper does not have enough length compared to its thickness. Now imagine that the sheet is 50 times longer (500,000 times longer than it is thick). After 6 folds, it is still 122 times longer than thick. In fact the sheet can be folded 9 times before reaching a similar thickness-length ratio to that of the letter-size one.

After some further calculations, Gallivan thought it would be possible to fold a piece of paper no fewer than 12 times, but there was one hitch. To achieve this feat, she would need to find a sheet a mile long. With admirable perseverance, Gallivan eventually discovered a long enough roll of paper and persuaded her parents to help her fold it inside a shopping mall. After seven hours she folded the paper for the eleventh time then posed for photos. Remarkably, there was still length enough for one more fold, just as she had predicted.

As Gallivan shows us, mathematical thinking is not just for mathematicians. In this chapter, I will show how everyone can benefit from such a mixture of precise yet imaginative reasoning. We will see how thinking in this way can help us to understand all kinds of complex real-world entities, from lotteries to voting systems. We will also explore a number of popular fallacies that employ math, and why they do not in fact add up. Finally, I discuss how anyone can improve his ability to think more carefully and avoid surprisingly common errors. We start with a look at some of the most important statistical terms and concepts, and how they can help make all of us better thinkers.

The Modern Stats Quo

The science fiction writer H. G. Wells boldly asserted, "Statistical thinking will one day be as necessary for efficient citizenship as the ability to read and write." Writing nearly a century ago, Wells's foresight is remarkable. After all, much of the information of the twenty-first century will be numerical. Not knowing how to understand such numbers will be as profound a handicap as is illiteracy.

American mathematician John Allen Paulos defines innumeracy, a term coined by cognitive scientist Douglas Hofstadter, as "an inability to deal comfortably with the fundamental notions of number and chance." Paulos helped to popularize the concept in his 1989 book of the same name, arguing that this inability is the result of poor early math education and a culture in which many people take a perverse pride in their numerical ignorance. Innumeracy matters, he insists, because it harms a person's general ability to think and reason carefully and to make good decisions.

Take the following example, an anecdote described by the sociologist Joel Best in his book, *Damned Lies and Statistics*. In 1995, Best attended a thesis defense in which the candidate argued that the number of young people killed or injured by firearms had doubled every year since 1950, citing a scholarly article in support of his contention. The problem is that this numerical "fact" is anything but. Even assuming there was only a single firearm-related incident in 1950, annual doubling would mean there were 2 such incidents the following year, 4 the year after, 8 the year after that, and so on. In 1965, according to the candidate's statistic, over 32,000 young people would have been killed or injured by firearms—a figure far higher than the total number of gun-related incidents reported that year. If we continue doubling, we find that by 1980 there would have been a billion young people killed or injured—four times the entire population of the United States. In 1987, the number killed or injured would have exceeded the total number of people who have ever lived on planet earth.

Such stories of statistical confusion are far from the exception. Many people struggle to understand the kinds of huge numbers often used by the media, scientists, and politicians. Quantifying very big numbers with simple visual analogies is one effective way of getting around this problem. For example, 100,000 is equivalent to the number of words in a good-size novel; a million then is equivalent to the number of words in ten such novels, and a billion equivalent to the number of words contained in 100 bookcases (each holding 100 novels). Another way of analogizing such numbers is to use time: if you spoke a word every second, it would take a little over a day to say 100,000 words; about a week and a half to say a million words, and almost thirty-two years to utter your billionth word.

Let's now look at some of the most frequently employed statistical terms, starting with the "mean," "median," and "mode." Imagine, for example, that we ask ten people at random for their age and receive the following responses: 7, 13, 19, 27, 27, 48, 51, 60, 75, and 83. The mean age of the group is found by adding all of these values together, then dividing this total by the number of results: $7 + 13 + 19 + 27 + 27 + 48 + 51 + 60 + 75 + 83 = 410/10 = 41$. The median is the middle value in the list (with the values arranged in increasing order). When—as in our list of ages—there are an even total of values, we add the middle two together and divide this figure in half: $27 + 48 = 75/2 = 37.5$. The mode is the number that recurs most frequently in the list. In the above example, the age 27 appears twice, while the others appear only once: hence the mode is 27.

As you can see, the mean, median, and mode values of a sample can vary quite significantly from one another, and these differences can sometimes prove deceptive. For example, imagine reading a company advertisement for a new job opening that states that its average salary is $5,000 per month. You apply and get the job. A month later you open your pay packet and find that you have only been paid $2,000. Furious, you march into the boss's office, only to be told that the average salary really is $5,000 per month. Look, says the boss, I employ 9 people: the 4 junior members of staff earn $2,000 per month, the 3 senior members earn $4,000 per month, and the 2 executive advisors earn $6,000 per month. My salary is $18,000 per month: (4 x

2,000) + (3 x 4,000) + (2 x 6,000) + 18,000 = 50,000, which divided by 10 is $5,000, just as I said in my advert. Though technically correct, it would have been more honest for the boss to have advertised using the median value ($4,000), or better still the mode ($2,000).

Sampling is another important concept in statistics. If you want to know how a country's population will vote in a forthcoming election, you could try asking every eligible voter, but for most countries this would take too long and cost too much. The way around such problems is to ask a representative cross sample of the voting population. However, working out whether a sample is truly representative or not is not easy—regardless of its size—as can be seen in the following classic story.

Literary Digest was a popular and prestigious American news periodical that ran surveys in presidential election years (beginning in the 1920s) in order to predict which candidate would prevail. The *Digest*'s predictions proved accurate for four elections running: in 1932, their prediction was even within 1 percent of the actual result. These predictions were made by sending out huge quantities of straw ballots to their readers and the general public: 20 million in 1932 (of which 3 million were completed and returned). In the 1936 election, based on the 2.3 million straw votes returned (out of 10 million sent out) the magazine projected that the Republican candidate, Alfred Mossman Landon, would win. George Gallup, a young psychologist who had interviewed 4,500 voters, disagreed and predicted that the incumbent president, Franklin Delano Roosevelt, would be reelected. Roosevelt did indeed win, with 60.8 percent of the vote to Landon's 36.6 percent—one of the biggest landslides in American history.

What went wrong with the *Digest*'s prediction? While enormous, their sample was biased: many of the ballot recipients were the *Digest*'s own readers, while others were selected randomly from phone books. This meant that they overselected people who were wealthy and more likely to vote Republican, because the people they sampled subscribed to a fairly conservative publication or had the means, in 1936, to have a telephone. On the other hand Gallup's survey, while being much, much smaller, was also more representative and therefore proved to be far more accurate.

Statistical sampling was an unlikely source of controversy during the 2000 United States Census, a national count of the American population that takes place once every ten years. In response to problems with its previous count, which had missed an estimated eight million people (a disproportionate number of whom were minorities, immigrants, and the poor), the national census bureau planned to introduce an element of sampling into its counting procedures. Though sampling was mostly supported by statisticians, Republicans opposed it, arguing that sampling was unconstitutional. The straight head count method went unchanged as a result of the dispute, leaving an estimated three million people still uncounted this time around.

Why such a fuss over the statistical method used for counting the national population? Because the accuracy of these counts matters enormously. The census is used to calculate the number of seats each state receives in the House of Representatives, as well as the proportion of some $200 billion a year in federal grants disseminated to states and localities. Big cities, such as Chicago and Los Angeles, have lost hundreds of millions of dollars because of previous undercounts. President Bill Clinton, a supporter of the bureau's sampling proposal, summed up the weaknesses of the nonsampling census model, arguing that it "distorts our understanding of the needs of our people and . . . diminishes the quality of life not only for them but for the rest of us, as well."

Two further terms often encountered in statistical discussions are "correlation" and "causation" (we came across these, briefly, in chapter 2). To recap, statisticians say that two sets of data are correlated when they appear to be linked in some way or dependent on one another. For example, the numbers giving a person's height and weight are correlated: the taller a person is, the heavier he will be. It is important to emphasize, however, that establishing a correlation does not necessarily prove that one value causes the other. If a scientific study shows that children who eat breakfast every morning before school obtain better grades, it might be because eating breakfast helps children to study and therefore achieve higher marks. But it might just as likely indicate the reverse, that getting better grades in school results in less stress and therefore higher appetite. The study's link might also be due to some other cause that affects both a child's appetite and his grades, or simply be entirely accidental.

The mathematical concept of probability is another that comes up in all kinds of situations. Having at least a basic understanding of it is important because its results are often surprisingly counterintuitive. Let's look first at the most famous and expensive mistake caused by a misunderstanding of what is and is not probable in a given situation: the Gambler's Fallacy. Imagine that you are in a casino and betting on the roulette wheel. Your luck is particularly bad and you wind up losing eight times in a row. Are you more likely to win on the ninth attempt than you were on the first, fourth, or eighth? Though many might be tempted to say yes, the answer is no, for the simple reason that the odds of a particular event in a random sequence (such as getting red or black on a roulette wheel) are totally independent of any previous ones. Thus the probability of getting red or black in roulette is always 50–50, regardless of any previous results.

Probability was a topic of discussion at a recent dinner I attended with friends and their guests. Someone at the table mentioned that they knew a family of ten children, all boys, and wondered aloud what kind of astronomical odds would account for such an oddity. I pointed out that the probability that each pregnancy will result in a boy or girl is, obviously, 1 in 2 and therefore the odds that a family of 10 children would result in

10 boys (or 10 girls) is equivalent to 2 x 2 x 2 x 2 x 2 x 2 x 2 x 2 x 2 x 2 or 1 in 1,024—a large number for sure, but hardly astronomical. Later that evening the same guest asked me whether the probability of the same family consisting of 5 boys and 5 girls would be 50 percent; the actual probability is in fact much lower: 252/1024 or 24.6 percent.

Such a discrepancy between intuition and the actual probability can be better illustrated if we take a more restricted example. In a family of 4 children, what is the probability that 2 will be boys and 2 girls? First, we calculate the total number of possible outcomes: 2 (boy or girl) x 2 x 2 x 2 (for four children) = 16. Thus, the odds of the children being all boys or all girls are each 1 in 16. The odds that only one of the four children is a boy or girl are each 4 in 16, as in (for one boy and three girls): BGGG, GBGG, GGBG, or GGGB. So there is a 2 in 16 chance that the four children will be all boys or all girls, and an 8 in 16 chance that only one of the four will be either a boy or a girl. That leaves a 6 in 16 (37.5 percent) chance for the remaining possible permutation of 2 boys and 2 girls.

A well-known example of probability's counterintuitive results is the Birthday Problem, which goes something like this: You are at a party with 22 other people. What is the probability that any two people present (excluding twins!) share the same birthday? We start by figuring the probability of a shared birthday for a party with just two people: the odds against the two birthdays matching are 364/365 multiplied by 100, more than 99.7 percent. For three people, the odds are: (364/365) x (363/365) x 100, almost 99.2 percent against any among them sharing a birthday. Adding a fourth person makes the odds narrow slightly further still to: (364/365) x (363/365) x (362/365) x 100 = 98.4 percent against. The probability still appears vanishingly small, but as we continue to add partygoers, the odds of finding a matching birthday starts to nudge up more and more quickly. For 10 people, the chance against any match of birthdays drops to around 88 percent, while for 20 people the odds against fall to around 59 percent. In fact, the probability that two people will have a matching birthday in a room of 23 people is greater than 1 in 2, around 50.7 percent.

As we can see, calculating the probability of an event will often provide a more accurate evaluation of a claim than intuition alone. For example, many people would be impressed by the achievements of a general who had fought and won six consecutive battles. Does this impressive-sounding record mean the general is a brilliant tactician of war? Not necessarily. Assuming roughly equal numbers of soldiers and equipment on each side, the odds of winning or losing a battle are 1 in 2—therefore the probability of winning six consecutive battles is equivalent to 2 x 2 x 2 x 2 x 2 x 2, or 1 in 64. There have been hundreds of generals throughout military history, meaning that—according to probability—a lucky few may well have achieved impressive strings of victories by chance alone.

According special meaning to purely coincidental events is a frequent and widespread error, caused by a misunderstanding of an event's probability. Have you ever thought of someone only to hear the phone ring and discover that he or she is on the other end of the line? The chances are that you have, and though an entertaining experience, it is not in fact an especially improbable one. This becomes obvious when you stop to consider the hundreds of people you know—vaguely or otherwise—and the correspondingly high number of thoughts you will likely have about them from day to day.

Mindful of these kinds of coincidences, the mathematician John Littlewood once calculated that the average person could expect to experience such a "miracle" once per month during the course of his life. Littlewood arrived at this astonishing conclusion by first defining a miracle as something considered highly significant and having a probability of one in a million. He then assumed that a typical person is particularly active about eight hours per day, during which time he sees and hears things occurring at a rate of about one per second. This amounts to around 30,000 events per day, or about a million each month. If any one of these million events is considered significant, among the vast number that are not, then we have a miracle.

Jackpots and Ballot Boxes

Evelyn Adams has to be one of the luckiest people on record. During a four-month period between 1985 and 1986, the New Jersey woman won the lottery not once but twice, collecting a total of $5.4 million. Unsurprisingly, the newspapers were astonished by the odds of such an event, which they calculated to be 1 in 17 trillion. Though correct, this figure refers to the likelihood that a specific individual buying a single ticket for exactly two New Jersey lotteries will win both times. In fact, Mrs. Adams—like many lottery players—bought multiple tickets, including hundreds at a time after her first win.

A better question to ask would therefore be: what is the probability of someone, from among the millions of American lottery players, hitting the jackpot twice in his or her lifetime? Harvard statisticians Percy Diaconis and Frederick Mosteller calculated the answer to this question: a better than 50 percent chance in seven years of a double lottery winner somewhere in the United States. Diaconis and Mosteller further calculated the odds of someone winning the lottery twice within a four-month period—as had Adams—as being around 1 in 30.

Sadly, Adams's wins did not bring her happiness; today she lives in a trailer, her money all gone. Adams claimed in an interview that she was constantly pestered for help following her double scoop, and that she suffered from a gambling problem. "Winning the lottery," Adams muses, "isn't always what it's cracked up to be."

I have never played the lottery in my life and never will. I agree with the French philosopher Voltaire, who famously described lotteries as a "tax on stupidity" (though it is, more specifically, one on innumeracy). The odds against winning the lottery are simply enormous—so large in fact that most people find it hard to appreciate the relevance of their size.

Here then is the math: in a game with 49 numbers to choose from, the probability of winning the lottery is equivalent to 6/49 (because there are 6 possible selections out of 49 that could match any one of your chosen numbers) x 5/48 (because there are 5 remaining possible matches and 48 balls) x 4/47 (4 possible matches and 47 balls) x 3/46 x 2/45 x 1/44 = 1 in 13,983,816 or almost 1 in 14 million. To make sense of such enormous odds, here is a comparison: the odds of flipping a coin and getting 24 heads in a row are a little over 1 in 16 million; the odds of dying from a bee sting are 1 in 6 million, while the odds of dying from a lightning bolt are 1 in 2 million. You are seven times more likely to be struck dead by lightning than to win the jackpot!

There are other reasons to avoid playing the lottery. Studies show that the poorest social groups spend much higher relative amounts on the lottery, making it highly regressive. This finding makes sense when we consider that the jackpot's allure is much greater for low-income individuals than for those who are already financially comfortable. Figures show that the average lottery player spends $313 per year on the lottery, whereas those earning less than $10,000 spend nearly twice that, an average of $597 per year. This is a terrible waste of money for the vast majority of players—the average "return" for long-term lottery play has been calculated at less than 50 percent; in other words, the average player will get back in winnings fewer than 50 cents for every dollar they put in. Low-income players who will put in $15,000 over twenty-five years can only expect to get back around $7,000 in winnings. In contrast, putting the same $597 a year into a compound interest bank account (at an interest rate of 4 percent) over twenty-five years will net a return of almost $26,000—nearly double the original amount put in, and almost four times the average lottery "winnings."

If—after all that you have read here—you still wish to continue playing the lottery, then I have some mathematical advice for you: how you pick your numbers will affect how much you stand to win (though, alas, it will not improve your chances of winning). The principle is simple: the more people who win the same jackpot, the smaller the winnings for each. In the United Kingdom draw of November 14, 1995, for example, no fewer than 133 people won a share of the £16 million jackpot—each winning just £120,000 as a result.

Therefore, when it comes to choosing your numbers, it is worth trying to avoid the combinations that other players might pick. This way, if you do win, you can be surer that you will win big. Researchers at Southampton University in the United Kingdom analyzed the number of winning tickets that shared the jackpot each week, comparing the total number of winners with the winning numbers. Their findings showed that picking unpopular numbers—such as 26, 34, 44, 46, 47, and 49—significantly reduces the chance of sharing the top prize with other players. The number 7 was shown to be the one especially to avoid—being played by more players than any other number—while a good pick was found to be the number 46, which was the least popular. Other advice includes avoiding sequences, such as 1, 2, 3, 4, 5, 6 (which, incredibly, is selected by 10,000 people each week in the United Kingdom), and limiting the numbers chosen below 31, because many players use birthdays to pick their numbers.

Another example of thinking mathematically comes from the hullabaloo surrounding the 2000 presidential election between Democrat Al Gore and Republican George W. Bush. The race proved an extraordinarily close one, ending with both candidates winning 48 percent of the national vote; Bush winning (according to CNN) 50,456,169 votes to Gore's 50,996,116. Even though Gore won the popular vote, he lost the election to Bush because the Republican secured more electoral college votes (271 to 266). The result was greeted by howls of outrage by many of Gore's supporters who argued that it showed that the electoral college system was undemocratic and urged it to be scrapped. I strongly disagree and will explain why.

First, a short refresher course on how U.S. elections work: In presidential elections, the winning candidate is the one who secures more than half the electoral college's 538 votes. The college's votes are decided between the states proportionate to their population, so that the larger states—such as New York, California, and Texas—have more votes than the less populous ones, such as Rhode Island or Wyoming. The winner of the electoral college votes for each state is decided by elections in each across the country on Election Day, with the vast majority of states awarding all of their college votes to the candidate who wins the most votes for that state. The victorious candidate is therefore generally the one with the broadest popular appeal across the many diverse states, big and small.

The electoral college system was established by the nation's founding fathers, as a compromise between the alternatives of electing a president by Congress or by the popular vote. Many of the "fathers" feared the "mob," believing that a large electorate is especially vulnerable to "hearsay" and "tumultuous passions." James Madison, the chief architect of the college, argued that a system of smaller elections in preference to a single national one would help protect minorities against the massed will of the majority: "A well-constructed Union," he asserted, must "break and control the violence of faction." In a single national election, candidates representing the majority would theoretically have no need to appeal to minorities and could ride roughshod over them. On the other hand, a national election composed of several smaller elections requires the candidates to appeal to a broad constituency if they are to have a chance of winning. The Madisonian system, by forcing the majority to seek the consent of the minority, has prevented the kind of factional wars that have plagued other countries and helped make the United States a more truly democratic nation, not less.

Massachusetts Institute of Technology professor Alan Nata-poff has used mathematics to come to the defense of the electoral college system against its ciritcs, such as the American Bar Association, which argues that it is "ambiguous" and "archaic." Natapoff's argument goes something like this: in fair elections, a voter's power derives from the probability that his or her vote might decide the election's outcome. Using mathematical modeling of how elections typically behave, Natapoff concluded that voters have a greater chance of turning an election when their vote is funneled through districts than when pooled together into a single national election. This is because a person's vote has a greater chance of deciding the outcome of their state, which in turn decides the national election, than it does of deciding a single national election.

Natapoff rejects the idea that the principle of every vote being equal means simply running every national election on a "one person, one vote" basis without subdividing voters into smaller local elections, arguing that it is too simplistic; after all, voters living under a dictatorship have equal (zero) voting power. True democracy, he contends, comes from giving "every voter the largest equal share of national voting power possible."

The math bears out his argument: imagine being part of a nation made up of 5 voters—the probability of your vote deciding the election's outcome depends on the other 4 voters splitting 2–2 between candidates A and B, an outcome with a probability of 37.5 percent. As the nation's size increases, individual voter power decreases. In a larger nation of 135 voters, for example, there are many more ways that the 134 other voters could split in such a way as to prevent your vote from deciding the race (e.g., 66–68 or 101–33) so that the probability of your vote making the difference drops to just 6.9 percent. This result, though, depends on the election being dead even, which real-life contests almost never are. In the same 135-voter nation, if voter preference for candidate A (or B) is 55 percent, the probability of your vote being decisive plummets below 0.4 percent.

Of course, elections are never decided by voters flipping a coin to choose between two candidates. Rather, they are always lopsided to varying degrees, with large blocs of voters having a preference for one candidate over another. Such lopsidedness is bad news for each individual voter, as the probability of his or her vote being decisive falls as the size of the electorate grows. Natapoff uses the analogy of a lopsided coin to illustrate this point: if you want to achieve an equal number of heads and tails, you will want to throw the coin as few times as possible—the more flips, the greater the coin's lopsidedness will manifest itself. In the same way, breaking big national elections into a series of smaller contests helps compensate against elections' natural unevenness, giving each individual voter a greater chance of casting the winning vote.

Critics of Natapoff's argument point to the 1888 election when the more popular candidate, Grover Cleveland (winning 48.6 percent of the national vote) was nudged out by his opponent Benjamin Harrison (who won 47.9 percent) with the electoral college votes splitting 233 to 168 in Harrison's favor. Natapoff responds that such rare results are a small price to pay to protect each individual's voting power with the threat that a handful of votes in one state or another could turn an entire election. Four years after his pipsqueaking defeat, Cleveland came back to beat Harrison under the very same rules as before.

Still, critics might point to the 2000 contest between Al Gore and George W. Bush where, they argue, Gore lost the election—even though he polled half a million more votes than his rival—because he lost the state of Florida by 537 votes. Canceling out a national vote majority of half a million with a single state's margin of a few hundred is unfair and undemocratic, they suggest.

There are two responses to this argument. The first is that the 2000 election was unusually close. Only three presidential elections since 1824 (the first election to record the popular vote)—1876, 1888, and 2000—have resulted in an electoral college winner who did not also attain a plurality of the popular vote. Only six elections (1880, 1884, 1888, 1960, 1968, 2000) out of forty-six (between 1824 and 2004) have finished with the two main candidates within 1 percent of each other. In comparison, the average margin of victory in presidential elections since 1824 has been around 9.5 percent. The bottom line: the 2000 election was an exception, making it a poor basis on which to argue against the existing electoral system.

The second response is that the 2000 election was not only extremely close in Florida (2,912,790 votes for Bush versus 2,912,253 for Gore) but in a number of other states as well, such as Iowa (638,517 votes for Gore versus 634,373 votes for Bush); New Hampshire (273,559 votes for Bush versus 266,348 for Gore); New Mexico (286,783 votes for Gore versus 286,417 for Bush); Oregon (720,342 votes for Gore versus 713,577 votes for Bush), and Wisconsin (1,242,987 votes for Gore versus 1,237,279 votes for Bush). Though Bush narrowly lost the overall national vote, he won many more states than Gore (30 to 21), a fact which helped tip the election in his favor.

For those seeking to reform the current electoral system, Natapoff has a couple of suggestions that he believes would boost voter turnout, force candidates to campaign even in states they are already likely to win or lose (which they generally ignore for that reason at present), and dissuade cheating in states where the result is extremely close. To help boost turnout, especially in states that are generally considered "locked up" by one party or another, Natapoff suggests changing the way the number of electoral college votes for each state is determined—from state population size to total voter turnout. At first blush this proposal sounds strange—why would a Democrat bother to vote in a Republican stronghold like Texas if he knew that his vote would help swell the state's potential number of electoral college votes for his candidate's opponent? Natapoff believes his system would force candidates to move toward the political center ground, avoiding extreme views or "preaching-to-the-choir" politics that typically dampen voter turnout—a far more significant problem for any democratic system. He uses the analogy of a poker player with a winning hand trying to keep the other players in the game in order to win a bigger pot.

Natapoff's other suggestion for reform is a response to the extremely close vote in Florida in 2000, which was marred by reports of cheating. Under this proposal, candidates who win by such a narrow margin would pick up only a proportion of the state's electoral college votes (the exact share would be worked out using a formula drawn from the shape of the bell curve), so that winning all of them would require a clear majority of votes cast.

If experts like Natapoff are right about the electoral college's advantages over a single national election, why then are so many Americans in favor of scrapping it—around 75 percent, according to opinion polls? Perhaps for some of the same reasons that many people play the lottery in exchange for "prizes" that are almost always, statistically speaking, going to be worth less than half the money they have put in over the years. People often act irrationally, regardless of the logical or mathematical arguments made for or against something. Let us look more closely at why this might be.

Why People Believe Weird Things

American author Michael Shermer literally wrote the book on why people believe "weird things." He defines these in four ways: a claim rejected by most experts, logically impossible, highly improbable, or one for which the evidence is limited or anecdotal. Shermer is quick to point out that such beliefs have little to do with a person's intelligence; smart and educated individuals can be just as vulnerable to believing "weird things" as anyone else.

One of the biggest reasons why people might accept an inferior idea in favor of a better one—such as supporting the idea of a single national election over the electoral college system—is a natural preference for simplicity. Given the choice between an idea that is simple to express and understand, and one that is more complicated and requires greater effort to comprehend, most people will naturally gravitate towards the simpler option, regardless of the two ideas' relative merits.

Such a strong bias towards simplicity is understandable in a world of immense and often baffling complexity. Many people feel they do not have the time to make the effort needed to engage more sophisticated ideas or arguments when simpler ones exist. Others tend to the belief that if an idea is a good one, it will be a simple one—a version of Ockham's razor (a philosophical position that contends that given a choice between a number of explanations or ideas, the individual should always prefer the simplest among them).

This interpretation, though, is a misunderstanding of what Ockham's razor actually means. Named after William Ockham (1285–1349), a Franciscan monk and the most influential philosopher of his time, the principle is most relevant when comparing two or more ideas of roughly equal credibility or explicatory power—giving the edge to whichever is the most concise or coherent. This point is well demonstrated when we consider that many excellent ideas are hugely complicated—such as Einstein's Theory of Relativity—while most poor or bogus ones are extremely simple, such as the principle of homeopathy ("like cures like"). The bottom line: do not reject an idea just because it is more complicated than another, especially if it is supported by good scientific or mathematical evidence.

What about the lottery players who continue to spend hundreds of dollars year after year, even when they typically win so little back? One argument is that many players know they are unlikely to win, but enjoy the thrill of taking part. According to this view, the lottery is more about hope than easy money. Another is sheer habit—lotteries take place every week, sometimes twice per week, and many players get into the cycle of playing their numbers on a particular day, between their other daily activities. The pattern of playing regularly leads to a further explanation for why many players continue indefinitely, even in the absence of any tangible results: fear. The possibility that a person's numbers might come up after he has decided to stop playing encourages individuals to keep on putting their money in.

An explanation for the continued participation of low-income individuals in lotteries is that it offers a potential "way out" of poverty or deprivation for a lucky few, that would never exist but for the individuals' willingness to take part, in spite of the odds. I have problems with this viewpoint: mostly because it ignores the vast majority of poor people who do not win the lottery, who only get poorer as a result of playing.

All of which brings me to a final reason why some people do the "weird thing" of playing the lottery, which is the dearth of clear or compelling alternatives. Poverty has often been described by social commentators as a trap, and for good reason—lacking the financial resources to live your own life has a pernicious effect on the individual and his family, affecting their health, culture, and imagination, as well as one's material well-being. Many players on low incomes do not see any other chance of escaping this trap, so they continue playing, even if they sense that it is unlikely to make any significant positive difference in their lives.

My advice to any players in such a situation would be to think of their money as an investment in tangible future possibilities, rather than in almost certainly unrealizable pipe dreams. Placing just a few hundred dollars a year into a compound interest savings account instead of playing it on the lottery is a good start. Think of all the ways that you could use the money to help enlarge your horizons and enhance your prospects: education is one very good investment, such as enrolling in an evening college, as would be putting the money towards a car (for those who do not have one) to improve your mobility.

Passivity seems to be especially common in people who persist in believing "weird things": from an unwillingness to think through a more complex idea because of the ready presence of a simpler one, to a continuation of bad habits, such as playing the lottery, because the alternatives are not clear or compelling in the same way as a long-shot fantasy (like winning the jackpot). The fact is that careful, logical thinking requires effort and does not come naturally. As Alfred Mander puts it in his book *Logic for the Millions*, "Thinking is skilled work. It is not true that we are naturally endowed with the ability to think clearly and logically—without learning how, or without practicing. People with untrained minds should no more expect to think clearly and logically than people who have never learned and never practiced can expect to find themselves good carpenters, golfers, bridge players or pianists." Regular practice in thinking for oneself is certainly more work than not doing so, but the price for not doing so may be infinitely higher.

Populations, Predictions, and Patterns

Mathematical thinking, like any other kind, can sometimes go wrong. From the misuse of statistics to selective reasoning and the misunderstanding of complex entities, this form of faulty thinking is surprisingly common. This is probably at least in part due to the additional allure that attaches to claims that employ numbers in their defense. Bear in mind, however, that these numbers do not always add up.

A good example of this is the oft-repeated assertion that our world is dangerously overpopulated. Such a pessimistic idea is in fact nothing new; people have worried about population growth throughout history. For example, the Greek playwright Euripides (c. 480–406 BC) wrote that the Trojan War was a divine act to, "purge the earth of an insolent abundance of people." Many Puritans sailed to the New World in the early seventeenth century because they considered England overcrowded.

The idea that there could be such a thing as too many people was formally expounded by the British economist Thomas Robert Malthus in his most famous work, "An Essay on the Principle of Population," published in 1798. Malthus argued that humans naturally reproduce at an exponential rate (2, 4, 8, 16, 32, 64, and so on), while the means of subsistence only increases arithmetically (2, 4, 6, 8, 10, 12, and so on). From this he deduced that populations had to be kept in check in order to avoid mass starvation. For the wealthy, like himself, Malthus recommended "moral restraint" as a method for stemming the growing tide of births. He was much more pessimistic, however, concerning the poor, believing that famine among them was both natural and inevitable.

Malthus's predictions failed in large part because he did not anticipate the coming agricultural revolution, which helped food production to meet—and even exceed—the needs of the world's growing population. It also helped widen access to prosperity by significantly reducing the price of staple foods. Rather than inevitably increasing as the population continues to expand, the incidence of famine has actually fallen sharply in the modern era.

Modern anxiety over the environmental damage caused by mass agriculture has replaced the Malthusian focus on famine. In 1968, the entomologist Paul R. Ehrlich authored *The Population Bomb*, in which he argued that the growing world population was endangering the very future of the Earth: "We must take action to reverse the deterioration of our environment before population pressure permanently ruins our planet. The birth rate must be brought into balance with the death rate or mankind will breed itself into oblivion. We can no longer afford merely to treat the symptoms of the cancer of population growth; the cancer itself must be cut out. Population control is the only answer."

Are human beings bad for the planet? I do not think so. Most people want to live in a clean and unpolluted environment, and many have even dedicated their lives to studying its climates and preserving its ecologies. The environmentalist Tony Juniper has argued that it is the profligate consumption of a rich few in the developed world, rather than the many, which is contributing to problems such as global warming. Technological advances, such as in renewable energy and a fairer distribution of resources, promise to help us all help the environment even more in the future.

The mathematical arguments used in favor of "population control" are the weakest, however. A statistic often given is that the number of people around the world has quadrupled in the past hundred years—from 1.6 billion in 1900 to 6.7 billion today. The implication is that this rate of growth will continue, and perhaps even accelerate further. Yet this argument ignores the actual population data showing that birth rates have fallen in every region of the world and nearly halved globally in the past fifty years. Various countries are even experiencing declining populations, including Russia, Germany, and Japan. Italy's population is expected to fall nearly 30 percent by 2050. In addition, exponential population growth simply cannot continue indefinitely. The United Nations Population Division has had to regularly revise its estimates downward since its 1968 prediction that world population will reach 12 billion by 2050. The agency now predicts a total of 9 billion, while others put the likely figure at nearer 7.9 billion.

Another argument for the world's "overpopulation" states that many countries are seriously overcrowded. Though it is true that large, developing countries such as China and Pakistan have population densities several times greater than the world average, several small and developed nations have even higher rates, including Belgium and the Netherlands. The fact is there is plenty of space for everyone. For example, the world contains around 57 million square miles of land, yet the entire global population could enjoy a high standard of living on just 6.5 million square miles (assuming a population density similar to that of the Dutch).

Increasing population levels are not due to a societal lack of control or education. Rather they are the result of technological and medical breakthroughs that have improved all our lives: clean water, vaccines, and antibiotics among others. More people are surviving childhood and living longer lives than ever before. At the same time, the standard of living for many has increased enormously. Though there is still much more work to be done to help the world's poorest, we should nonetheless celebrate the achievement of science and human ingenuity that these facts represent, not mourn it.

A further example of mathematical thinking gone awry is known as the "Jeane Dixon effect," named after the late American "psychic." Dixon is best known for supposedly predicting the assassination of President John F. Kennedy in a magazine interview seven years before the event—except that she did not. What Dixon actually said is: "It [the 1960 Presidential election] will be dominated by labor and won by a Democrat. But he will be assassinated or die in office though not necessarily in his first term."

But let's give Dixon the benefit of the doubt and consider her vague prediction a "hit." Her supporters point out that the probability that Kennedy would die in office was not high (never mind that Dixon did not mention any Democrat by name in her prediction). How then do we explain this famous example of seeming clairvoyance? By looking at all of the available data, not just the few that show positive correlations. For example, Dixon later contradicted herself by predicting that Nixon, not Kennedy, would win the 1960 election. In fact, she made thousands of predictions during a forty-year soothsaying career, most of them completely wrong. Among her many, many "misses" were her predictions that World War III would begin in 1958, that there would be a cure for cancer in 1967, and that the Soviets would be first to land a man on the moon.

A version of the "Dixon effect" can be demonstrated by the following example: imagine a stock market "expert" who sends annual messages to his clients, predicting whether the stock market will rise or fall in the year ahead. The expert sends a thousand of these messages, half of them stating that the market will rise and the other half that it will fall. The following year he does the same for the five hundred clients who received the correct prediction. He continues like this, sending out his messages each year, so that after five years there are around thirty clients who have received the correct predictions five years in a row. To these thirty individuals, the expert's foresight will appear uncanny.

Scientists can be just as vulnerable to this effect as can other people. This is because they are trained to search results methodically for patterns, yet it is far from easy to tell the difference between a pattern that is meaningful and one that has occurred randomly. In his book *Fooled by Randomness*, Nassim Nicholas Taleb gives the example of "cancer clusters" (a greater than expected number of cancer cases in a particular area) to illustrate this problem. Taleb points out that "randomness does not look random," arguing that such clusters will often occur entirely by chance. For example, imagine throwing sixteen darts randomly at a board. Each dart has an equal probability of hitting any place on the board. If we divide the board into sixteen equally sized parts, we might expect one of the darts to end up in each of them. However, such an evenly spaced result would occur only rarely. Performing this experiment over many different boards, we will find that many have more than one dart in some parts and none in others. Many seeming patterns, therefore, are nothing of the kind. Telling a real pattern from coincidence requires the reader to look at all the data that does not fit his theory, as well as all that does.

A further related problem for scientists is the "publication bias"—the tendency to publish research with a positive outcome more frequently than research that does not show any significant results. This bias is important because it risks causing the subject under investigation to be misrepresented by incomplete data. A 2005 study published in *Nature* reported that 6 percent of scientists admitted to rejecting data because the information contradicted their previous research. Fifteen percent also reported ignoring observations because they had a "gut feeling" that they were inaccurate.

This bias is also seen in our television news programs, which often prefer the drama or controversy of "bad news" stories to good ones. Little wonder then that the casual viewer comes to the familiar conclusion that the world is going to hell in a hand basket. Take heart, however, in the fact that what we see on our screens is only a very selective snippet of the day's events. Do not let your optimism for the world disappear in a news flash.

A final example of misguided mathematical thinking, known as "intelligent design," argues that the universe's striking orderliness is evidence against the possibility that its features evolved over time. The assumption, however, that impressive-looking patterns could not arise spontaneously from a large enough set of data is incorrect.

For instance, the mathematical constant Pi (3.141 . . .) contains an endless sequence of digits, with each digit as likely to occur at any given point in the sequence as any other. Yet within this morass of numbers are plenty of examples of seemingly meaningful combinations. A famous example of this is the sequence "999999," which occurs as early as position 762 after the decimal point, even though its probability is literally one in a million. Another example is "12345678," which occurs more than 186 million digits after the decimal. We can even find the first 8 digits of Pi (including the 3) within Pi, starting at position 50,366,472. Because Pi is infinite, we could in theory go on like this indefinitely, finding every possible length and combination of digits that we care to think of.

A more sophisticated demonstration that patterns will always arise from a sufficiently large set of random objects comes from a branch of pure mathematics known as Ramsey Theory, named after the mathematician Frank P. Ramsey. Researchers using this theory have been able to demonstrate the existence of numerous mathematical patterns in sets of numbers that emerge without any conscious design. To show this, try writing out a series using the letters A and B without creating any sequences of three evenly spaced A's or B's. For example, if we randomly jot down the series BABB we must next write A to avoid three successive B's, making BABBA. We then add another A to reach: BABBAA, after which we cannot go on without unintentionally creating an orderly sequence for both letters. If we add another A, we have three successive A's; if instead we add a B to the series, we find that we have three evenly spaced B's, at positions 1, 4, and 7. In fact, it has been proven that any combination of nine A's and B's will always contain such a pattern. All of which shows, as the mathematician Theodore Motzkin once noted, that complete disorder is simply impossible.

Logic: The Science of Good Thinking

Lewis Carroll, author of *Alice in Wonderland*, was also a mathematician and an inventor of logic puzzles. Carroll argued that such puzzles helped the mind to develop "the habit of arranging your ideas in an orderly and get-at-able form . . . the power to detect fallacies, and to tear to pieces the flimsy illogical arguments, which you will so continually encounter in books, newspapers, in speeches and even in sermons."

Bearing Carroll's advice in mind, let's start the final section of this chapter with a small exercise of our logic skills using one of his puzzles:

My saucepans are the only things I have that are made of tin.
I find all your presents very useful.
None of my saucepans are of the slightest use.

What is the ultimate logical implication we can draw from these three statements? First, let's clarify what the statements are, using a simple notation for each:

S: It is my saucepans.
T: It is made of tin.
P: It is your presents.
U: It is very useful.

We can then rewrite the three sentences in the puzzle using our notation, including the contrapositive (the reverse) for each:

S → T (not T → not S)
P → U (not U → not P)
Not S → U (not U → S)

Now we examine our symbolic statements until we find a way to string them together to reach a conclusion (note that we can read the statements backwards as well as forwards), as in: P → U → not S → not T. That is, the second statement plus the third statement (read backwards), plus the first statement's contrapositive (read backwards).

Finally, we translate the implication of this symbolic sequence (the first and final parts) into ordinary language: "Your presents are not made of tin."

The Concise Oxford English Dictionary defines logic as "the science of reasoning, proof, thinking, or inference." Logic helps us to analyze a person's thinking—our own included—and evaluate whether it is likely correct or not. Of course logic has its limits—choosing a potential mate or professing a religious faith are common activities performed on the basis of a personal evaluation that may or may not be strictly rational (the details of which are beyond the scope of this book). In many situations throughout our lives, however, logic enhances and enriches our ability to think and reason clearly, and to avoid the trap of self-deception.

Logic is used to determine the relations between different statements and to obtain conclusions from true ones. A statement that is either true or false is known as a "proposition"—an example is "Dieting is good for your health." Modern technology means that we are constantly bombarded with all kinds of propositions—from gossip and urban myths to advertising slogans and media stories. Knowing how to evaluate such statements as accurate or otherwise is an important part of thinking for yourself.

When presented with a proposition, ask yourself what are the reasons or assumptions—technically known as premises—that support it. For example, a person who hears that going on a diet will make him healthier might assume it is likely true because its premises appear plausible. These can be listed as:

Diets restrict calorie intake.
Consuming fewer calories helps the dieter reach a healthy
   weight.
Overweight people are vulnerable to various health problems.

If we translate these statements, as we did for the Lewis Carroll puzzle, we have:

D: It is a diet.
R: It restricts calorie intake.
O: It is overweight.
P: It is vulnerable to various health problems.

D → R (not R → not D)
R → not O (O → not R)
O → P (not P → not O)

Examining these premises, we reach the conclusions that, "Diets help one to not be vulnerable to various health problems" (D → R → not O → not P), and, "Not dieting makes you vulnerable to various health problems" (not D → not R → O

→ P). But we now need to look at these premises more carefully to determine whether they are actually valid. If any one of the premises supporting an argument is invalid, then the entire argument falls.

It is true that people who are overweight are more vulnerable to various health problems than those who are not. And it is also the case that eating fewer calories will (in most cases) help an overweight person eventually reach a normal weight. The first premise in our list above, however, is problematic, because there is good evidence that diets do not necessarily restrict the intake of calories.

This is because dieting involves the deliberate control of eating habits, which warps the individual's relationship with food. Rather than simply eating when hungry, the dieter chooses what and when and how much to eat according to a tightly controlled plan. As a result, the consuming of food is no longer governed by biology. Diets replace our natural eating controls (such as eating less in the evening if we have overeaten at lunch) with the precise segmentation of meals, which tells the dieter how many calories are "allowed" for breakfast, lunch, and dinner. Such restrictions do not fit easily with the often social nature of eating (with family or at a restaurant, for example), making it all too easy for individuals to end up "breaking" their diet. Doing so causes dieters distress, which perversely makes them eat more (while nondieters generally respond to distress and the reduced appetite that often results by naturally eating less). Considering these facts, we conclude that the proposition is false: diets are more bad than good for your health.

There are all kinds of propositions. For example, they can be either positive ("London is in England") or negative ("London is not in France"). They can have one subject ("Humans are animals") or several ("Humans, bees, and elephants are animals"). We can connect two or more propositions together using the words "and" and "or" ("John is a man and Joan is a woman," "John is a man or John is a woman"). As we saw in the Lewis Carroll puzzle above, a statement is equivalent to its reverse ("A triangle has three sides" and "It does not have three sides, it is not a triangle"). We assemble these or other kinds of propositions together to form arguments whose conclusions are either valid or invalid.

Bogus propositions (those that are misleading or irrelevant) can often be tricky to spot, so our attention now turns to how to recognize them. A technical flaw that makes an argument or piece of reasoning unsound or invalid is known as a logical fallacy. Dauntingly, there are many possible fallacies—far too many, in fact, to list them all here. Nonetheless, acquaintance with the most common and important, as given below, will stand any thinker in good stead.

One common form of logical fallacy distracts the individual from alternative options. A classic example of this is the False Dilemma, such as: "You are either part of the problem or part of the solution." Such black-and-white terms prohibit cautious or nuanced thinking in favor of knee-jerk reactions.

Another class of logical fallacies involves appealing to emotions rather than to the particular merits of an idea or argument. An example of this is the "Appeal to Consequences": "Evolution cannot be true, because if it were, it would mean we are no different from monkeys."

"Ad hominem" (Latin for "to the man") is another way of avoiding the actual details of a given piece of reasoning, by criticizing the person making the argument rather than the argument itself: "Mr. Smith is a well-known member of the Flat Earth Society, so we can safely assume that anything he says is probably nonsense."

Even when a critic addresses a particular argument or idea, it is often caricatured, rather than considered in its original or best form—a fallacy known as the "Straw Man." Example: "Environmentalism is nothing but a bunch of tree-hugging hippies trying to hold back industrial progress."

Many arguments are made using analogies—saying that

something has a particular property because something else sufficiently similar has that same property. Beware, however, of comparisons that are inaccurate, which fall under the logical fallacy known as a "False Analogy": "Atheists are like the Bolsheviks—just as the Bolsheviks had no belief in a God and absolute moral standards, so are atheists prone to anarchism and immorality."

As we saw in the earlier examples of the "Dixon effect," many people will mistakenly see meaningful links between events where none in fact exist. "Post Hoc Ergo Propter Hoc" (Latin for "after it therefore because of it") is an example of this kind of causal fallacy, occurring when the thinker assumes that because one thing came after another the earlier one caused it. For example: "The president did not wear a cowboy hat during his trip to Texas, and later that year he lost Texas in the general election. If only he had worn that hat!"

Some logical errors are caused by category fallacies, believing that a whole of something will necessarily possess the same qualities as a part or vice versa: "One of her sons is always getting into trouble, so they must all be a bad lot," and "The Catholic Church is very conservative; therefore all Catholics are very conservative."

Perhaps the most important logical errors to avoid are those caused by not being clear with the definitions that we use. This is because careful and effective reasoning depends on precise definitions, neither too broad ("A dog is an animal with four legs and a tail"; but so is a cat, and many other animals) nor too narrow ("A poem consists of verses that rhyme"; but there are other kinds of poetry that do not use rhyme). Sometimes a definition will include the term being defined as a part of the definition—a fallacy known as a "Circular Definition": "A God is a being with divine characteristics" ("divine" is an adjective meaning "godlike," so this definition is circular).

In his final book, *The Demon-Haunted World*, astronomer Carl Sagan writes at length about the "fine art of baloney detection." Concerned with the spread of superstitious thinking at the tail end of the twentieth century, he argues that true reasoning should lead to a conclusion based on a clear and precise train of thought, rather than a conclusion that is selected simply because it is liked.

Sagan explains his lifelong love affair with science as the product of its fascinating nature—a combination of two seeming opposites: "an almost complete openness to all ideas, no matter how bizarre . . . a propensity to wonder . . . but at the same time . . . the most vigorous and uncompromising skepticism, because the vast majority of ideas are simply wrong."

It is this delicate balance—between an openness to the beauty and wonder of novel ideas and ways of seeing and understanding our world, and a capacity to pause, to analyze, to question and, quite often, to doubt—that is at the heart of thinking well. Logic—often seen, mistakenly in my view, as cold and calculating—need not detach anyone from the mystery of love and faith, the ambiguities inherent in living a human life. Rather, careful reasoning and independent thought help to keep our feet on the ground—from where we have the best view of the stars above us.

10

The Future of the Mind

The end of our journey together is in sight, as we conclude with a look ahead to what might await us, and future generations of minds, in the years that follow. In an age of accelerating medical and technological advances, futurist promises are becoming ever more ambitious and grandiose: from magnetic energy caps that will give savant skills to everyone, to the ability to download whole books in our heads, and even the eventual merging of man with machine. As intriguing as these claims are, I wonder if such prognostications are especially accurate, and more to the point whether they are even desirable. In these closing pages I examine the evidence presented for these arguments and give my own view of what the future might bring to our thinking, feeling, imagining selves.

The "Genius Cap"

We met Professor Allan Snyder of the University of Sydney's Centre for the Mind, and his TMS (transcranial magnetic stimulation) cap device for inducing savant skills in "normal" people, back in chapters 1 and 5. To summarize: Snyder's cap is placed on an individual's head and sends safe level bursts of magnetic energy to the frontal lobes for around a quarter hour. As a result of this treatment, the individual reportedly can replicate certain savantlike skills, though the effect wears off after a few hours. If this is true, technological advances in the near future might make savants of everyone.

Snyder believes that the brain is constantly computing the raw sensory data it receives, though in most people the results of these unconscious computations are lost as the conscious mind pulls the details together to form generalized patterns, shapes, and concepts. Savants, on the other hand—according to Snyder—do not conceptualize in this way and so are able to draw on these calculations by the brain; it is this privileged access that explains savant abilities. The professor theorizes that this same access could be unlocked in non-savants by temporarily "shutting down" activity in the left frontal hemisphere of the brain (which deals with logical and conceptual thought), allowing savantlike abilities to briefly emerge.

This theory that savant talents arise in this way is problematic because it fails to account for the special creativity of high-functioning savants. It might explain, for example, why and how Stephen Wiltshire can draw city landscapes in minute and exact detail, but not how the French savant Gilles Trehin draws similarly intricate drawings using nothing but his imagination. It might help explain how some savants possess perfect pitch—assuming it derives from some mental processing that instantly and unconsciously dissects sound in the brain—but not how the teenage autistic savant Matt Savage has composed several albums of original jazz music.

Most tricky for Snyder's theory are savant mathematical skills such as mine. He suggests that they might arise from an as-yet-unknown fundamental mechanism in the brain that instantly equipartitions numbers (breaks them down into smaller equal parts). In contrast, my own theory, described in chapter 5 and based on my own savant experience, is that such skills are the result of hyperconnectivity within the brain. I see no reason why other savant skills could not similarly arise from such hypercon-nectedness, especially as researchers know it occurs in conditions such as autism.

Even worse for Snyder's idea, however, is the experimental evidence that appears to contradict it. In 2000, Adelaide doctors Robyn Young and Michael Ridding tested the hypothesis using applications of TMS to the fronto-temporal lobes of seventeen volunteers. The scientists reported that in only two was any significant short-lived improvement in artistic or memory ability noted. Young and Ridding conclude that, "these findings suggest that these skills may be limited to a small percentage of the 'normal' population just as they appear to be in the disabled population." Snyder's own studies with small numbers of subjects show similar temporary changes in certain skills in only a small proportion of them.

I should also note that the level of ability following TMS application is not especially striking: Snyder's studies to date indicate some improvement in some subjects in tasks such as drawing everyday animals (horses or cats) from memory, proofreading short texts with duplicated words, and—as discussed in chapter 5—estimating the number of dots flashed up briefly on a screen. In contrast, no post-TMS subject has demonstrated the ability to draw beautiful, complex pictures or compose a novel piece of music, factorize a four-digit number, or learn the grammar of a foreign language (or invent one of his own). It appears then that TMS is only a very imperfect way of tapping the capacities of the savant mind, the biology of which is much more complex and subtle than Snyder allows for.

Many neuroscientists agree that TMS could likely never be used in the way envisaged by Snyder and his colleagues. Eric Wassermann, a neurologist at the National Institute of Health's Neurology Institute, has been testing TMS on hundreds of research subjects over many years and points out that not one of them has ever "revealed a sudden genius for anything." Darold Treffert, the world's leading researcher on savant syndrome, is also skeptical: "The likelihood of significant savant abilities emerging in a ten- or twenty-minute TMS session in normal volunteers is, in my view, zero."

Even if TMS cannot be shown to produce anything like savant skills in most people, it remains a technology with much promise in the future treatment of neurological conditions such as epilepsy and schizophrenia. A 1999 study by researchers at the University of Göttingen in Germany reported promising results from the use of repetitive TMS in nine patients with epilepsy that was unresponsive to drugs. All but one of the patients showed significant reduction in the frequency and severity of their seizures following daily TMS therapy. The treatment has also shown results in the reduction of auditory hallucinations, a common problem for those suffering from schizophrenia. What is more, these results have been shown to last for weeks at a time; in one patient the improvement lasted two months.

Medical Mind Boosting

TMS therapy is not the only attempt by researchers to help give people's minds a boost: some are working towards the creation of "smart pills"—drugs designed to improve memory and mental ability. U.S. companies such as Helicon and Sention are developing medications to treat patients with brain disease and injury, but they also foresee a near future where many healthy people will make use of such drugs to give themselves a "brain lift." By way of examples, they point to the already current use, by some travelers combating jet lag, of modafinil, a drug normally prescribed in the treatment of daytime drowsiness in narcolepsy patients. They also point to students who take Ritalin—used to treat attention deficit–hyperactivity disorder—to help them cram for exams.

The potential market for such brain-boosting drugs is huge: in the United States alone there are four million patients with

Alzheimer's disease, and another twelve million with a condition known as mild cognitive impairment (which can often lead to Alzheimer's). Tens of millions more have what scientists term "age-associated memory impairment"—a mild form of forgetfulness. Annual U.S. sales of dietary supplements believed to improve memory and cognitive function—from vitamin B_{12} to ginkgo biloba—exceed $1 billion, even though the scientific evidence that they work is scanty at best.

Not surprisingly, some scientists have raised concerns about the possible use of medicinal treatments as "lifestyle" drugs. Moral philosopher Leon R. Kass, head of the President's Council on Bioethics, has argued that the use of such treatments would cheapen any achievements resulting from their use. If we consider excellence to be the product of effort, talent, and discipline alone, swallowing a pill is tantamount to cheating.

Smart pills are supposed to work either by increasing blood flow to the brain or by boosting the levels of one or other of the neurotransmitters thought to play a role in learning and memory. Their testing has been primarily on mice and rats, to see whether they run through a maze faster after being given a particular drug. According to British neuroscientist Steven Rose, a critic of much of the research into "smart drugs," such tests tell us very little, if anything, about a drug's potential effect on the human brain. For example, a drug might stimulate the mouse's appetite, which would make it run faster through the maze to receive its food reward; these successful trials would more likely be the result of an animal's increased motivation rather than any heightened brain power.

Rose also points out that among the few studies on human patients, who were often suffering with Alzheimer's, average samples are very small (fewer than ten people per study), and the effects of the drugs are often assessed subjectively by doctors and nurses, rather than by repeated trials with different patients. Many of these patients also suffer from anxiety, anger, or depression, and Rose suggests that any memory improvement might simply be the result of the drug alleviating such negative feelings.

Such skepticism of the often overblown claims made for smart pills is further supported by the findings of a 1998 study by researcher Nancy Jo Wesensten and colleagues at the Walter Reed Army Institute of Research. Wesensten asked 50 volunteers to stay awake for 54 hours without sleep, giving them a placebo, 600 milligrams of caffeine (equivalent to six cups of coffee) or one of three doses of modafinil (100, 200, or 400 milligrams) after 40 hours. The subjects were subsequently given a series of cognitive function tests to assess their performance. The researchers found that both caffeine and the highest dose of modafinil helped restore cognitive performance to normal levels—in other words, coffee worked just as well as the smart drug.

Other researchers are concerned about the long-term safety of many of these drugs, the effects of whose chemicals on neurological function and behavior are still unknown and their potential side effects considerable. Take the example of the relatively humble nicotine patch, which researchers at Duke University discovered helps improve cognitive function in some Alzheimer's patients, as well as adults with ADHD and schizophrenia. The patch's side effects, however, including elevated heart rate and blood pressure, insomnia, nausea, and dizziness, were considered too great to recommend its use in the treatment of these conditions.

Of course, boosting performance with the use of drugs is nothing new—from the morning cup of coffee before work, to the age-old consumption across cultures of alcohol, tobacco, and peyote. Amphetamines, usually considered a modern "upper," were first synthesized more than a century ago. In a 1987 study of the fifty-one largest orchestras in the United States, it was found that a quarter of musicians used beta-blockers—originally developed and prescribed to treat high blood pressure— to help them control their stage fright. Today, though, the sheer range and variety of drugs promising to improve attention span, heighten memory, raise efficiency, and boost intellectual performance are staggering. Even Eric Kandel, a leading figure in the development of these drugs who shared the 2000 Nobel Prize in medicine, is appalled by the idea that college students could be popping his pills to improve their grades. The drugs he is working on, he insists, are for those suffering from serious illness only, such as memory loss from cancer chemotherapy or Alzheimer's. For healthy people, according to Kandel, they are simply not worth the risk.

Brains Beyond Biology

Science fiction's tall tales became a little shorter just before the new millennium with the story of Johnny Ray—a fiftyish man from Georgia who suffered a brain-stem stroke in 1997, after which he could not speak or move. A year later neuroscientist Philip Kennedy fitted an electronic chip inside Ray's head, with which he was able to detect signals within his brain and decode them using a computer. In tests, Ray was asked to think about specific motions, such as moving his arms. Kennedy took the corresponding brain signal and programmed it to move a cursor. Over time, Ray learned to move the cursor by himself and type messages, using nothing more than the power of thought. Eventually he was able to move the cursor in the same way that most people might move their hand or turn their head, fluently and unself-consciously—as though the cursor had become an extension of himself. Ray had, according to Kennedy, become, "the world's first cyborg."

Kennedy foresees other possible future technologies that interface our minds directly with computers, through which we would be able to download books, search the Web, play games, control robots remotely, and even drive cars just by thinking. He is not alone, either: researchers in Rhode Island have taught monkeys how to play a computer pinball game with their thoughts rather than their paws, while in Australia scientists have developed a device that lets the wearer turn on lights and radios using mental commands.

Other researchers are even more enthusiastic about the future of such mind-machine interface technologies, arguing that they will eventually help humans transcend their own biology and merge with computers. British robotics professor Kevin Warwick is one of the most prominent proponents of this view, arguing that humans need to be "upgraded" if we are to have any hope of preventing intelligent machines from one day overthrowing us. Warwick attracted headlines in 2002 when he attempted a form of primitive telepathy between himself and his wife Irena. The couple were wired up with electrodes inserted into their arms' median nerves, from which electro-chemical impulses could be converted by a computer into digital signals and sent between one another across the internet. As his wife made simple finger movements, Warwick reported a tingling sensation in his arm: "like a mild electric shock." The result is just the beginning, he believes, predicting that a generation from now people will communicate their thoughts telepathically to one another through chip implants inserted in their brains.

Such claims have brought considerable criticism, however, from Warwick's own academic peers, such as Inman Harvey, a professor of cognitive and computing sciences, who believes Warwick is either a "charlatan" or a "buffoon," and calls his predictions "ridiculous." Others have pointed out that Warwick's tingling sensation could have been achieved just as well with a pair of vibrating cell phones. Warwick may have become the victim of his own enthusiasm for his peculiar beliefs, turning him more into a showman than a scientist.

Inventor and artificial intelligence pioneer Ray Kurzweil is an even more flamboyant and higher-profile evangelist for the eventual merging of man and machine. The creator of the world's first computer-based reading device for the blind, Kurzweil is also the author of several books propounding his belief that humans will soon have the option of digitizing our personalities and "uploading" them into computers, achieving a kind of virtual-reality immortality.

As with Warwick, Kurzweil's predictions have been criticized by scientists as naïve and unworkable. Both are based on the mistaken idea that the human mind is analogous to a digital computer. For one thing, every brain is different—not just by virtue of biological variations in sex, age, and health, but also because our minds are constantly changing in reaction to our environment—external and internal. Every thought, every daydream, every emotion alters the brain's fantastically intricate structure in subtle but definite ways. As a result, many neuroscientists doubt anything like human upgrading or uploading will ever be possible, no matter how far advanced the technology becomes.

Michael Chorost, author of the memoir *Rebuilt: How Becoming Part Computer Made Me More Human,* is a good summary of technology's achievements, and of its limitations. Born with severe hearing loss, Chorost suddenly went completely deaf in his midthirties. His memoir recounts how he was fitted with and adapted to life with a cochlear implant, noting that, though a remarkable invention, it is nonetheless a very poor simulation of the natural human auditory system. The implant's sound quality is mediocre, especially in noisy situations, meaning that the wearer still has to resort occasionally to lip reading and guesswork to understand what someone might be saying to him. Some people are not able to use artificial cochleas at all, though nobody knows why this might be.

Chorost is singularly unimpressed by the claims and predictions of researchers like Warwick and Kurzweil, rejecting them as "over-weening techno-optimism." Computer hardware is not going to replace our brain's "wetware" anytime soon, he insists;

his implant made him feel more human, not less, helping him to connect more deeply and intimately with other people. Chorost also illustrates the fundamental difference between a human mind and a computer from his unique experience: even with his implant, he found that feelings and belief affected how much of a conversation he understood. Human communication, he realized, was far more than simply input (hearing) and output (speaking).

The philosopher Michael Polanyi's view of how we acquire knowledge and understanding jibes with Chorost's experience. Polanyi argued that human knowledge is fundamentally complex, personal, and embodied—the antithesis of a computer's data. For him, the human body is "the ultimate instrument of all our external knowledge, whether intellectual or practical . . . experience [is] always in terms of the world to which we are attending from our body."

Cognitive scientists Andy Clark and David Chalmers emphasize the essential role bodily activity plays in human cognition. Children, they point out, learn by interacting with the world around them far more than by simply reading books and listening to words. Similarly, numerous educational methods encourage children to use tangible objects, such as beads and blocks, to help them acquire an intuitive understanding of abstract mathematical concepts.

Clark and Chalmers also point to various studies showing that gesturing helps speakers to structure their thoughts as they attempt to verbalize them. University of Chicago psychology professor Susan Goldin-Meadow provides accounts of some of these studies in her book, *Hearing Gesture: How Our Hands Help Us Think*. In one of these, she describes how children participating in a science lesson on seasonal change used gestures to facilitate their understanding of a problem:

Gail was trying to decide where the sun would shine most directly. She was representing the sun with her left fist, which she held at a distance from the globe, at the same time that she said the sun was shining, "right about straight on the equator." She then gestured without speaking—she traced a line from her fist (her sun hand) to the globe. After finishing her gesture, which landed just below the equator on the globe, she revised her statement as follows: "No, right about *here*. More on the . . . southern hemisphere. Shining directly about over, somewhere over here."

A further example of how our bodies help to facilitate and structure our thought comes from the game of Scrabble. Players commonly manipulate their letter tiles before placing them on the board to stimulate and direct their mental search for possible words. In this way the tiles can be said to have become a part of the player's cognition. Similarly, as I type this book's words on my computer screen, I am continuously adding or cutting some, and rearranging or reworking others. These actions are a vital part of a writer's thinking process, stimulating or developing new ideas while refining or modifying others.

Perhaps the biggest advantage that an embodied cognition confers on us, though, is the capacity to give meaning to our lives and selves through committed choices. Our experience of decision making in the physical world is shaped by uncertainty and awareness of our bodies' vulnerability. The philosopher Hubert Dreyfus argues that this helps us to be careful in the choices that we make, and take them seriously, knowing that each will have potentially significant consequences for our future.

Like Chorost, Polanyi, Dreyfus, and those working in the field of embodied cognition, I am appalled at the idea of "upgrading" my humanity, even considering its many imperfections. Like the scientists, I doubt whether a machine could ever replicate something as rich and subtle as human intelligence. The fantastic visions of Warwick, Kurzweil, and others seem a response to the perceived hopelessness of the human condition, motivated more by a science of despair rather than one of imagination as they claim. Such visions of engineered nirvanas are the product of angst, not ambition.

A Humane Future

Let us look ahead then to a different future to that envisioned by the cyber-evangelists, one where advances in medicine and technology will continue to improve our lives, but also where the boundary between man and machine remains unblurred. What might such a future look like?

Perhaps not so far in the distance, new treatments will intelligently and successfully treat serious, often devastating neurological conditions, such as epilepsy, schizophrenia, Alzheimer's disease, autism, and clinical depression. For example, researchers have begun to exploit virtual reality technology to help high-functioning autistic children learn a range of crucial life skills. Researchers Justine Cassell and Andrea Tartaro of Northwestern University studied six children with high-functioning autism aged between seven and eleven as the children interacted during an hour-long session with a real-life child, and with a "virtual peer" called Sam.

Sam has moppy brown hair and wears a light blue T-shirt with darker blue trousers. Designed to appear around eight years old and of indeterminate sex, he is projected on a large screen in front of the child and engages him in conversation and play. Unlike real-life children, Sam never gets tired or loses his patience. The researchers can vary his appearance and conversation so that the children are exposed to different forms of behavior. The goal is to help the children practice basic social skills, such as joining a game or holding a conversation, so that they can prepare for such interactions in the real world. Early studies show that the children respond well to their virtual friend, producing more two-way conversation when they speak with Sam than with real-life children.

In Haifa, Israel, researchers are also using virtual reality to help autistic children learn how to cross the street, a necessary skill that many struggle with and where real-world learning is often prohibitively dangerous. Six autistic children aged seven through twelve used virtual simulations programmed by researcher Yuval Naveh over a one-month period, in which they learned how to cross virtual streets, to wait for the virtual green light at the crosswalk, and to look left and right for any virtual cars. The children showed significant improvement throughout the learning process, and were subsequently able to transfer the skills learned to real-life road crossing situations.

Other neurological conditions are also slowly yielding to ingenious medical advances. In Alzheimer's research, a vaccine has been developed that clears plaque formations from the arteries in the brain, believed to be a key cause of the disease's effects. The vaccine works by stimulating the body's immune system to produce antibodies that attack the protein in the plaques. Scientists hope that a combination of vaccine and other therapies treating the neuron damage and cognitive decline characteristic of the condition will offer significant relief to future patients.

A cutting-edge technology that helps doctors diagnose patients with either bipolar disorder or schizophrenia is the "lifeShirt"—a computerized vest that continuously monitors the patient's movements and collects data on physiological measures such as respiration and heart rate. From this data, researchers found that those with bipolar disorder were hyperactive and roamed widely around a new environment, whereas those with schizophrenia tended to move far less in the same situations. Doctors often struggle to distinguish between the two conditions in diagnosing certain patients, so the data from the vest is highly helpful in ensuring that patients receive accurate and effective treatment for their symptoms.

Novel therapeutic technologies such as these and many others will be underlined in the years to come by increasingly more sophisticated scientific knowledge and understanding of how our brains work. One of the most interesting avenues of neurological exploration is functional magnetic resonance imaging (fMRI)—a scanning technique that detects changes in blood flow to particular areas of the brain, with which researchers identify and map the parts involved in particular mental processes.

Some of the most recent research (March 2008) that used fMRI scanning indicates that its measurements of brain activity contain much more information about underlying neural processes than previously realized. Scientists have started to produce impressive results by using computer models to analyze the data produced by such scans, assessing not only the regions of the brain that are activated during particular mental states, but also how this diffuse activity in the brain is ultimately fused together to form our everyday sense of perception.

In a study headed by neuroscientist Jack Gallant, of the University of California, Berkeley, researchers used fMRI scans to record the activity in the visual cortex of subjects while they looked at several thousand randomly chosen images. The cortex's neurons are known to respond to specific aspects of a visual scene, producing signature activities that the scientists compiled to develop a computer model that would predict the pattern of brain activity produced by a particular picture. When the volunteers were later shown a new image not included in the original set, the computer model was able to predict with 80 percent accuracy which image out of a thousand possibilities the person was looking at.

Gallant and his research team plan to use their technique to better understand the brain's visual system, by developing different computational models for various theories, and testing their ability to accurately interpret the data collected from brain scans. This approach could also be used to help scientists study cognitive activities that are otherwise difficult to analyze, such as attention, imagination, and dreaming.

Together with such impressive advances in medicine and technology, I hope for continuing progress in our cultures, too, particularly in the way that society views individuals with different minds. In the not-too-distant past, autistic savants were considered of little scientific or intellectual interest and often treated as mere curiosities or performing seals. Even to this day autistic savants are too often viewed as robots, or computers, freaks, or even supernaturally endowed—in short, anything but human. And yet, as I have argued elsewhere in this book, it is our humanity that makes our extraordinary abilities possible.

With all that we have begun to learn in recent decades about the intricacy and idiosyncrasy of "normal" brains and minds, and with the growing awareness of the wide variability in conditions as complex as the autistic spectrum, such distorting and hurtful misconceptions will—I hope—decline in the years ahead. Better still, society will find ways to make best use of the talents and energies of differently able minds, maximizing the depth and diversity of its intellectual capital in the face of the many challenges, and opportunities, that lie ahead for all of us.

The future need not belong to the futurists. Given the chance to contribute meaningfully within a truly inclusive meeting of every kind of mind, each one of us can use our brains to do what they have always done best: imagining a better and brighter tomorrow.

Acknowledgments

This book is the fruit of several years of scientific research and personal reflection. Throughout this time my mind and life have been immeasurably enriched by all kinds of shared ideas and insights, experiences and adventures. To the following I am particularly grateful for their help, encouragement, and/or for all that I have learned with them:

Many thanks to my editor at Free Press, Leslie Meredith, and her editorial assistant, Donna Loffredo. A special mention to my previous editor, Bruce Nichols, for his warmth and enthusiasm.

Also to my editors at Hodder, Rowena Webb and Helen Coyle.

A *merci* is also due to Catherine Meyer at Les Arenes for her kind and thoughtful input.

Thanks as ever are due to my literary agent, Andrew Lownie, whose friendliness and efficiency are second to none.

A big thank-you to all the scientists and researchers with whom I have had the great fortune to learn much about myself and how my mind (and every other) works. Special thanks to: Vilayanur Ramachandran, Shai Azoulai, Edward Hubbard, Bruce Miller, Darold Treffert, Simon Baron-Cohen, Julian Asher, Daniel Bor, Chris Ashwin, Jac Billington, Sally Wheelwright, Neil Smith, Gary Morgan, and Charles Leclerc.

Special thanks also to Jean-Philippe Tabet and Margo Flah for their generosity and hospitality during much of the writing of this book.

And finally, innumerable thanks as ever to my family and friends for all their love, support, and encouragement. In particular, *takk* to Sigríður Kristinsdóttir and Hallgrímur Helgi Helgason, Laufey Bjarnadóttir, and Torfi Magnússon, Valgerður Benediktsdóttir and Grímur Björnsson; *thanks* to Ian and Ana Williams and Olly and Ash Jeffery, and *merci* to Jérôme Tabet.

Bibliography

Ackerman, S. "Optical Illusions: Why Do We See the Way We Do?" *HHMI Bulletin,* June 2003.

Aitchison, Jean. *The Seeds of Speech: Language Origin and Evolution.* Cambridge: Cambridge University Press, 2000.

Amie, J. "A Bird's Eye View: Ultraviolet Vision Lets Birds See What Humans Can't." *Imprint Magazine,* 2007.

Andreasen, Nancy C. *The Creating Brain.* New York: Dana Press, 2005.

Asperger, Hans. *Die "Autistischen Psychopathen" im Kindesalter (Autistic Psychopathy of Childhood).* Archiv für Psychiatrie und Nervenkrankheiten, 1944.

Baillargeon, Normand. *A Short Course in Intellectual Self-Defense.* New York: Seven Stories Press, 2008.

Bains, S. "Mixed Feelings." *Wired,* issue 15.04, March 2007.

Baron-Cohen, S., D. Bor, J. Billington, J. Asher, S. Wheelwright, and C. Ashwin. "Savant Memory in a Man with Colour Form-Number Synaesthesia and Asperger Syndrome." *Journal of Consciousness Studies,* volume 14, 2007, pp. 237–251.

Best, Joel. *Damned Lies and Statistics.* Berkeley: University of California Press, 2001.

Bickerton, Derek. *Bastard Tongues: A Trail-Blazing Linguist Finds Clues to Our Common Humanity in the World's Lowliest Languages.* New York: Hill and Wang, 2008.

Blakeslee, S. "A Disease That Allowed Torrents of Creativity." *The New York Times,* April 8, 2008.

Blakeslee, S. "Mathematicians Prove That It's a Small World." *The New York Times,* June 16, 1998.

Bor, D., J. Billington, and S. Baron-Cohen. "Savant Memory for Digits in a Case of Synaesthesia and Asperger Syndrome Is Related to Hyperactivity in the Lateral Prefrontal Cortex." *Neurocase,* volume 13, 2008, pp. 311–319.

Boroditsky, Lera, Lauren A. Schmidt, and Phillips Webb. "Sex, Syntax, and Semantics" in *Language in Mind.* Cambridge, MA: MIT Press, 2003.

Bibliography

Brown, Donald E. *Human Universals.* Philadelphia: Temple University Press, 1991.

Brunvand, Jan Harold. *The Vanishing Hitchhiker: American Urban Legends and Their Meanings.* New York: W. W. Norton & Company, 1981.

Burger, Edward, and Michael Starbird. *The Heart of Mathematics: An Invitation to Effective Thinking.* Springer. 2005.

Butterworth, Brian. *The Mathematical Brain.* London: Macmillan, 1999.

Buzan, Tony. *Use Your Memory.* London: BBC Books; rev. ed., 1989.

Carey, B. "Anticipating the Future to 'See' the Present." *The New York Times,* June 10, 2008

Carroll, Lewis. *Alice's Adventures in Wonderland.* London: Macmillan, 1865.

Carroll, Lewis. *Symbolic Logic and the Game of Logic.* New York: Courier Dover Publications, 1958.

Carson, Shelley H., Jordan B. Peterson, and Daniel M Higgins. "Decreased Latent Inhibition Is Associated with Increased Creative Achievement in High-Functioning Individuals." *Journal of Personality and Social Psychology,* volume 85, number 3, 2003.

Chomsky, Noam. *Language and Mind.* New York: Harcourt Brace & World, 1968.

Chomsky, Noam. *Reflexions on Language.* New York: Pantheon Books, 1975.

Chorost, Michael. *Rebuilt.* Boston: Houghton Mifflin, 2005.

Damasio, Antonio. *Descartes' Error: Emotion, Reason, and the Human Brain.* New York: Grosset/Putnam, 1994.

Dehaene, Stanislas. *Numerical Cognition.* Oxford: Blackwell ed., 1993.

Dunbar, Robin. *Grooming, Gossip, and the Evolution of Language.* Cambridge, MA: Harvard University Press, 1996.

Ebbinghaus, Hermann. *Memory,* London: Thoemmes Continuum; English ed. (1913), 1998.

Ehrlich, Paul R. *The Population Bomb.* New York: Ballantine Books, 1968.

"Fatally Flawed: Refuting the Recent Study on Encyclopaedic Accuracy by the Journal *Nature.*" Encyclopedia Britannica, Inc., March 2006.

Flaherty, Alice Weaver. *The Midnight Disease.* Boston: Houghton Mifflin, 2004.

Flora, C. "The Grandmaster Experiment." Psychology Today, July/August 2005.

Fromkin, Victoria, and Rodman, Robert. *Introduction to Language.* New York: Harcourt Brace, 6th ed., 1997.

Gallivan, B. "How to Fold Paper in Half Twelve Times—An 'Impossible Challenge' Solved and Explained." The Historical Society of Pomona Valley.

Gallivan, Britney C. *How to Fold Paper in Half Twelve Times.* Pomona, NY: Historical Society of Pomona Valley, Inc., 2002.

Gardner, Howard. *Frames of Mind.* New York: Basic Books, 10th ed., 1983.

Gardner, Howard. *The Shattered Mind: The Person After Brain Damage.* New York: Knopf, 1975.

Bibliography

Garreau, J. "A Dose of Genius." *The Washington Post,* June 11, 2006.

Gilmore, C., S. McCarthy, and E. Spelke. "Symbolic Arithmetic Knowledge without Instruction." *Nature,* May 2007, pp. 589–91.

"Ginny and Gracie Go to School." *Time,* December 10, 1979.

Goldin-Meadow, Susan. *Hearing Gesture.* Cambridge, MA: Belknap Press of Harvard University Press, 2003.

Goleman, Daniel. *Emotional Intelligence—Why It Can Matter More Than IQ.* London: Bloomsbury, 1995.

Gombrich, Ernst H. *Art and Illusion.* New Jersey: Princeton University Press, 1961.

Gould, Stephen Jay. *The Mismeasure of Man.* New York: W. W. Norton & Company, 1981.

Greenberg, Joseph H. *Universals of Language.* Cambridge, MA: MIT Press, 1963.

Harris, S., S. Sheth, and M. Cohen, *Annals of Neurology.* volume 63, issue 2. December 10, 2007, pp. 141–147.

Hennacy, Ken, Peter Slezak, and Diane Powell. *All in The Mind* transcript. ABC Radio National, October 22, 2005.

Hermelin, Beate. *Bright Splinters of the Mind.* London: Jessica Kingsley Publishers, 2001.

Herrnstein, Richard, and Charles Murray. *The Bell Curve.* New York: Free Press, 1994.

Hively, W. "Math Against Tyranny." *Discover,* November 1996.

Hoffman, Donald D. *Visual Intelligence: How We Create What We See.* New York: W.W. Norton & Company, 1998.

Hostetter, Autumn B., Martha W. Alibali. "On the Tip of the Mind: Gesture as a Key to Conceptualization," in K. Forbus, D. Gentner, and T. Regier (eds.), Proceedings of the Twenty-Sixth Annual Conference of the Cognitive Science Society, 2004.

Howe, Catherine Q., and Dale Purves. *Perceiving Geometry.* New York: Springer-Verlag, 2005.

Howe, Michael J. A. *Genius Explained.* Cambridge: Cambridge University Press, 1999.

Johnson, G. "To Test a Powerful Computer, Play an Ancient Game." *The New York Times,* July 29, 1997.

Karl, H. S. Kim, Norman R. Relkin, Kyong-Min Lee, and Joy Hirsch. "Distinct Cortical Areas Associated with Native and Second Languages." *Nature,* number 388, 1997.

Klemmer, Scott R., Björn Hartmann, and Leila Takayama. "How Bodies Matter: Five Themes for Interaction Design," in *Proceedings of DIS06: Designing Interactive Systems: Processes, Practices, Methods, & Techniques,* 2006.

Lakoff, George. *Don't Think of an Elephant.* Vermont: Chelsea Green, 2004

Laplante, Eve. *Seized: Temporal Lobe Epilepsy As a Medical, Historical, and Artistic Phenomenon.* New York: HarperCollins, 1993.

Bibliography

Lehrer, J. "The Reinvention of the Self." *Seed Magazine,* February 23, 2006.

Lenneberg, Eric H. *Biological Foundations of Language.* New York: John Wiley & Sons, 1967.

Lönnrot, Elias (Kirby W. F., trans.). *The Kalevala.* London: Dent & Sons Ltd. 1966.

Luria, Alexander (L. Solotaroff, trans.). *The Mind of a Mnemonist.* Cambridge, MA: Harvard University Press, 1968.

Malthus, Thomas R. *An Essay on the Principle of Population.* Cambridge: Cambridge University Press, 2-volume ed., 1990.

Mander, Alfred E. Logic for the Millions. New York: Philosophical Library, 1947.

Martinson, Brian C., Melissa S. Anderson, and Raymond De Vries. "Scientists Behaving Badly." *Nature,* number 435, 2005.

Miller, G. "The Man Who Memorized Pi." *ScienceNOW,* 2005.

Miller, George A. "The Magical Number Seven, Plus or Minus Two." *Psychological Review,* 63, 1956.

Moyer, Robert S., and Landauer, Thomas K. "Time Required for Judgments of Numerical Inequity." *Nature,* number 215, 1967.

Nasar, Sylvia. *A Beautiful Mind.* New York, Simon & Schuster, 1998.

Noice, H., and T. Noice. "What Studies of Actors and Acting Can Tell Us About Memory and Cognitive Functioning." *Current Directions in Psychological Science,* volume 15, issue 1, April 2006, pp. 14–18.

Nunberg, Geoffrey. *Going Nucular.* New York: Public Affairs, 2004.

Olding, P. "The Genius Sperm Bank." *BBC News Magazine,* June 15, 2006.

Orwell, George. "Politics and the English Language." *Horizon,* number 76, 1946.

Osborne, L. "A Linguistic Big Bang." *The New York Times.* October 24, 1999.

Page, Scott. The Difference: *How the Power of Diversity Creates Better Groups, Firms, Schools and Societies.* Princeton, NJ: Princeton University Press, 2007.

Paulos, John Allen. *Innumeracy.* New York: Hill and Wang, 1988.

Penrose, Roger. *The Emperor's New Mind.* Oxford: Oxford University Press, 1989.

Pinker, Steven: *The Language Instinct.* New York: HarperCollins, 1994.

Polgár, László, and Endre Farkas. *Nevelj zsenit!* (Bring Up Genius!). Budapest: Interart, 1989.

Ramachandran, Vilayanur S. *Phantoms in the Brain.* London: Harper Perennial, 1999.

Ramachandran, Vilayanur S. *A Brief Tour of Human Consciousness.* New York: Pi Press, 2005.

Ramachandran, V. S., and W. Hirstein. "The Science of Art: A Neurological Theory of Aesthetic Experience." *Journal of Consciousness Studies,* volume 6, numbers 6–7, 1999.

Bibliography

Ravitch, Diane. *Language Police—How Pressure Groups Restrict What Students Learn.* New York: Alfred A. Knopf, 2003.

Reid Priedhorsky, Chen Jilin, Shyong K Lam, et al. "Creating, Destroying, and Restoring Value in Wikipedia." 2007 International Conference on Supporting Group Work, 2007.

Rix, J. "Painting? I Can't Turn It Off." *Times Online,* July 14, 2007.

Roberton, Lynn, and Noam Sagiv. *Synesthesia: Perspectives from Cognitive Neuroscience.* New York: Oxford University Press US, 2005.

Rochat, Philippe. *The Infant's World.* Cambridge, MA: Harvard University Press, 2001.

Rorschach, Hermann. *Psychodiagnostics Plates.* Bern: Hans Huber Publishers, 1921.

Rosenzweig, M. R. "Comparisons among Word-Association Responses in English, French, German, and Italian." *The American Journal of Psychology,* volume 74, number 3, September 1961, pp. 347–360.

Roszak, Theodore. *The Cult of Information.* Berkeley: University of California Press, 1994.

Sacks, Oliver. *The Man Who Mistook His Wife for a Hat.* London: Duckworth, 1985.

Sagan, Carl. *The Demon-Haunted World.* New York: Random House, 1996.

Saunders, Pearce and Amato. The Original Australian Test of Intelligence, 1983.

Schiff, S. "Know It All." The New Yorker, July 31, 2006.

Shenk, David. *Data Smog.* New York: HarperEdge, 1997.

Shermer, Michael. *Why People Believe Weird Things.* New York: W. H. Freeman & Company, 1997.

Silverstein, Michael. *Talking Politics.* Chicago: Prickly Paradigm Press, 2003.

Sokal, Alan D. "Transgressing the Boundaries: Towards a Transformative Hermeneutics of Quantum Gravity." *Social Text,* numbers 46–47, 1996.

Sternberg, Robert. *Beyond IQ. A Triarchic Theory of Intelligence.* Cambridge: Cambridge University Press, 1985.

Surowiecki, James. *The Wisdom of Crowds.* New York: Doubleday, 2004.

Taleb, Nassim Nicholas. *Fooled by Randomness.* New York: W. W. Norton & Company, 2001.

Tammet, Daniel. *Born on a Blue Day.* London: Hodder & Stoughton Ltd., New York: Free Press, 2006.

Treffert, Darold A. *Extraordinary People.* Backinprint.com, 2000.

Tulving, Endel. *Elements of Episodic Memory.* Oxford: Oxford University Press, 1985.

Vygotsky L. S. and A. R. Luria. *Studies on the History of Behavior: Ape, Primitive, and Child.* New Jersey: Lawrence Erlbaum, 1993.

Wakefield, Andrew J., S. H. Murch, A. Anthony, J. Linnell, et al. "Ileal-Lymphoid-Nodular Hyperplasia, Non-specific Colitis, and Pervasive Developmental Disorder in Children." *The Lancet,* volume 351, 1998.

Bibliography

Wales, Jimmy. "Special Report: Internet Encyclopaedias Go Head to Head." *Nature,* number 438, 2005.

Wynn, Karen. "Addition and Subtraction by Human Infants." *Nature,* number 358, 1992.

Yamaguchi, Makoto. "Questionable Aspects of Oliver Sacks' (1985) Report." *Journal of Autism and Developmental Disorders,* volume 37, number 7, 2006.

Yule, George. *The Study of Language.* Cambridge: Cambridge University Press, 1996.

Zhaoping, L., and Jingling, L. "Filling-in and Suppression of Visual Perception from Context—A Bayesian Account of Perceptual Biases by Contextual Influences." PloS Computational Biology, February 15, 2008.

Index

Index

Index

Darwin, Charles, 54, 94, 135, 155
Darwin, Erasmus, 155
Data Smog (Shenk), 213
Davidson, Richard, 17–18
"decay theory of forgetting," 82
decision making, 212–13, 263
Deep Blue computer, 28, 30
de Groot, Adriaan, 29
Dehaene, Stanislas, 24, 134
déjà vu phenomenon, 76–77
dementia, 84–85, 160–61
Demon-HauntedWorld,The (Sagan),
    251–52
depression, 11, 51, 86, 257, 264
    postpartum, 160
depth perception, 177
Descartes, René, 153–54
"devil's tuning fork" illusion, 183
Dewey decimal system, 216–18
Diaconis, Percy, 230
Dickinson, Emily, 5
Diderot, Denis, 204
dietary supplements, 257
direction, sense of, 15, 49
Down, J. Langden, 21
Dreyfus, Hubert, 263
Dryden, John, 157
Dunbar, Robin, 198

Ebbinghaus, Hermann, 70, 81
Ebbinghaus illusion, 173
Edelman, Gerald, 65
Ehrlich, Paul R., 241
Einstein, Albert, 2, 171, 205, 238
elaborative encoding, 65–68, 74, 85
elections, U.S., 225, 235–36, 243
    electoral college system in,
        232–37, 238
    of 2000, 232–33, 235–36, 237
electroencephalograph, 17–18
"11 Plus" exam, 42–43
email, 151, 213, 214–15, 216
embodied cognition, 262–63
Emotional Intelligence (EQ), 50–51
Emotional Intelligence (Goleman), 50
emotions, 50, 79, 186
    advertising in provoking of, 212

brain in stimulating and process-
    ing of, 10, 12, 188, 189, 212
learning ability as affected by,
    33–34
in logical fallacies, 250
memory reconstruction and, 59,
    60–62, 63, 64, 70, 77, 216
universal aesthetic principles in
    stimulation of, 186, 188, 189,
    190
Emperor's New Mind,The (Penrose),
    156
encyclopedias, 4, 204–8, 218
epilepsy, 8, 37, 106, 138, 139, 256,
    264
    creative thought as linked to,
        157–58, 171
episodic memory, 60–65
Ericsson, K. Anders, 54
Escher, M. C., 186
"Essay on the Principle of Popula-
    tion, An" (Malthus), 241
Euclid, 147
euphemisms, 194–95
Exiting Nirvana (Park), 169
eye, visual perception in, 2, 174–76,
    178, 183, 184, 186
eyewitness testimony, 78

facial recognition, 59–60
factorization, 143, 149
"False Analogy," 250–51
False Dilemma, 250
"false friends", in second languages,
    115–16
false memory phenomenon, 79–81
feelSpace belt, 15
Fei Xu, 127
fetus, brain of, 9
Feynman, Richard, 35, 219–20
field mode, in memory, 61–62
Fitzgerald, Michael, 171, 200
Flaherty, Alice, 160
Flynn, Suzanne, 106
fMRI (functional magnetic reso-
    nance imaging), 200–201
folklore, 101, 198, 201–3
Fooled by Randomness (Taleb), 244

Index

Index

Index

number instinct, *cont.*
  languages as providing evidence
    of, 128–29, 130–32
  manipulation of sums into shapes
    and patterns in, 142–45
  mental visualization of numbers
    in, 22–23, 57–58, 72–74, 125,
    133–36, 139, 141–45, 178
  numerical memory and, 57–58,
    59, 71–74
  occasional fallability of, 145
  prime numbers and, 23–25,
    143–46
  recognizing semantic relations and
    shapes in, 140–41, 142–44
  see also mathematical thinking
numbers, 125–52
  autistic savants as fascinated by,
    146–48
  infinity concept in, 146–48
  language and expression of, 101,
    128–29, 130–32
  semantic relations of, 140–41, 143
  see also mathematical thinking
Nunberg, Geoffrey, 195

observer mode, in memory, 61–62
Ockham, William, 238
Ockham's razor, 238
Ogden, Charles, 192
Ojemann, George, 105–6
Onfray, Michel, 36
onomatopoeia, 101, 113
"op (optical) art," 185–86
optical illusions, 4, 180, 181–86
oral literature, 70, 86, 201–3
Orwell, George, 191–93, 194

Paivio, Allan, 72
parietal cortex, 55
parietal lobes, 127, 139
Park, Clara Claiborne, 169
Parkinson's disease, 199
Pascual-Leone, Alvaro, 16–17
patterns, 49, 238–39
  faulty mathematical thinking on,
    244, 245–46
  in language and grammar, 87, 93

in prime numbers, 148–49
recognizing of, 28, 30, 33
Paulos, John Allen, 223
"peak shift," 187
Peek, Fran, 58
Peek, Kim, 58–59, 76, 78, 138
peer pressure, in advertising, 211
Penrose, Lionel, 182
Penrose, Roger, 156, 182
Penrose triangle illusion, 182, 186
Perceiving Geometry (Purves), 183
perception, 32, 38, 191, 266
  advertising tricks in, 211–12
  of bilinguals vs. monolinguals,
    106–7
  of chess grandmasters, 29
  see also visual perception
"perceptual focus effect," 211–12
perceptual illusions, *see* optical illusions
"perceptual problem solving,"
  aesthetic principle of, 188
Peterson, Jordan, 158–59
phantom limb phenomenon, 13–15
Phillips, Webb, 118
phonology, 107–10
photographic memory, 58, 59
Pi, 57–58, 72, 153, 245
Picasso, Pablo, 171, 188
Pick's disease, 127–28
pidgin language, 95, 162, 164
Pinker, Steven, 90, 197
pitch contour, of sentences, 108–9
Plato, 153, 203
poetry, poets, 37, 49, 101, 154, 159,
  187, 189
  by autistic individuals, 168–70
  in oral literature, 201, 202–3
Polanyi, Michael, 262, 263
Polgar sisters, 53, 54
politicians, 49, 194, 196
"Politics and the English Language"
  (Orwell), 192–93
polls, election, 225–26
Ponzo, Mario, 181
Population Bomb, The (Ehrlich),
  241–42
population growth, 240–43

About the Author

Daniel Tammet is a writer, linguist, and educator. A 2007 poll of 4,000 Britons named him as one of the world's "100 living geniuses." His website company, Optimnem, has provided foreign language instruction to thousands around the globe. His previous book, the *New York Times* bestseller *Born on a Blue Day: Inside the Extraordinary Mind of an Autistic Savant,* has been translated into eighteen languages. He lives in Avignon in the south of France.